# INTO THE EXTREME

# INTO THE EXTREME

## U.S. Environmental Systems and Politics beyond Earth

VALERIE OLSON

University of Minnesota Press | Minneapolis | London

Published by the University of Minnesota Press
111 Third Avenue South, Suite 290
Minneapolis, MN 55401–2520
http://www.upress.umn.edu

Printed in the United States of America on acid-free paper



24 23 22 21 20 19 18 10 9 8 7 6 5 4 3 2 1

Library of Congress Cataloging-in-Publication Data
Names: Olson, Valerie, author.
Title: Into the extreme : U.S. environmental systems and politics beyond Earth / Valerie Olson.
Description: Minneapolis : University of Minnesota Press, [2018] | Includes bibliographical references and index.
Identifiers: LCCN 2017056773 (print)| ISBN 978-1-5179-0254-4 (hc) | ISBN 978-1-5179-0255-1 (pb)
Subjects: LCSH: Space environment. | Space environment–Political aspects. | Extraterrestrial bases. | Architecture, Postmodern.
Classification: LCC TL1489 .047 2018 (print) | DDC 629.47/7–dc23
LC record available at https://lccn.loc.gov/2017056773

For Morrigan

# CONTENTS

# ABBREVIATIONS

| | |
|---|---|
| as-can | astronaut candidate |
| as-ho | astronaut hopeful |
| BIO-Plex | Bioregenerative Planetary Life Support Systems Test Complex |
| Constellation Program | NASA human spaceflight program (2005–2009) |
| HDC | Habitability Design Center (Johnson Space Center) |
| ISS | International Space Station |
| JSC | Johnson Space Center (Houston, Texas) |
| JPL | Jet Propulsion Laboratory (Pasadena, California) |
| NASA | National Aeronautics and Space Administration |
| NEO | near-Earth object (asteroids, comets, meteoroids, debris with Earth-crossing orbits) |
| NEEMO | NASA Extreme Environment Mission Operations (space analog program) |
| NOAA | National Oceanic and Atmospheric Administration |
| NSBRI | National Space Biomedical Research Institute (Houston, Texas) |
| PHO/PHA | potentially hazardous object / potentially hazardous asteroid |

# INTRODUCTION
## A SYSTEM OF SYSTEMS

As DAYS SHIFT INTO NIGHTS, social spaces appear differently. One such space to live, visible only at night, is an inhabited machine orbiting high above the Earth. Like any other satellite, the collaboratively built International Space Station shows up as a point of reflected sunlight transiting the glittery night sky. But unlike most satellites launched over the course of sixty years of spaceflight, this one houses a select gathering of people, animals, plants, and microbes. They interact inside conjoined capsules full of noisy devices and packed-up provisions that sustain them beyond Earth's living envelope. Flying at 27,000 kilometers per hour, the spacecraft stays aloft with bursts of fuel, but it is always in free fall. Speeding yet falling, the inhabitants and other contents of the station are weightless, suspended together in the crux of hyperartificially built and terrestrially unnatural conditions. The global partnership network responsible for the station promotes it as a laboratory uniquely removed from natural planetary forces.

But when I started fieldwork on human spaceflight programs, no one spoke of space's unnaturalness or naturalness. *Nature* is not the working term for outer space. Nor is the station an entirely extraordinary kind of thing. For thousands of engineers, scientists, technicians, and astronauts who work to maintain this spacecraft and its connections to Earth, the station is related to other things in the universe. In the technical

The International Space Station. Courtesy of NASA.

parlance shared by spaceflight practitioners across the globe, the station is a semiclosed ecological *system* containing other kinds of *systems* within a solar *system* situated within the *extreme environment* of outer space.

In 2007, I was inside a closed-up room in the United States that is organized to interact with the International Space Station. This is a spaceflight control room, secured within a National Aeronautics and Space Administration facility on the outskirts of Houston, Texas. As one of NASA's ten far-flung field centers, the Johnson Space Center and the many companies located nearby form the hub of U.S. civil human spaceflight. The flight control room I was in—known popularly as "mission control"—is an artificially lit and cooled spacecraft-like enclosure. These rooms and the "All systems go!" catchphrase associated with them have become almost banal elements in movies about spaceflight. But they are also iconic sites that extend U.S. systems science and technology into new spaces. I watched two huge screens at the front of the room: one showed a graphic of the station moving over a flattened global map, and the other fed live video from the station of two male U.S. and Russian crew members fixing part of a cooling system. The flight controllers who staff the control room in shifts take on the acronyms of the systems they manage. I sat between RIO and BME; I could see FLIGHT and CAPCOM. Jeff, a young white engineer training to be an ECLSS (Environmental Control and Life Support Systems) flight controller, explained his work to me:

> I describe myself as a systems engineer. . . . The one thing that makes up such a big part of our thought process at NASA is, and how we define how our projects evolve, is environment, environment, environment . . . as they say, you're in a bubble . . . everything interacts with everything. . . . Today NASA is looking more at the larger picture, the interactions of everything, whether it's a human system or a machine. And I think the way we do things is right in that you have your engineering disciplines and your space and life sciences disciplines, and they focus at this micro level on their area, but then you have system engineering integrating all those systems . . . bringing it all together.

As a life-support worker, Jeff-as-ECLSS is tasked with accounting for the "interactions of everything" across perceived spatial and material scales beyond his earthly experience: exhaled breath, molecules gassing off plastics, shedding skin cells, beads of urine, leaking hoses, dying plants, mutating bacteria, satellite debris, micrometeoroids from the solar system's edges, and radiation from the Big Bang. Enthralled with his training to work in that kind of relational space, Jeff told me that the extensive systemic interactions that spaceflight draws together stand "in contrast to pretty much any other project on Earth."

During my seven years of fieldwork, U.S. government civil servants and contractors showed me how astronautics work, in its simplest terms, is about managing systems in environmental conditions categorized as extreme. Such extremes are the stuff of both science and fiction. Hurtling around the planet, the station's crews of five or six do not experience days and nights. Astronauts described to me in vivid terms how they watched the slow glow and sharp blaze of sunrises and sunsets every forty-five minutes over an electric-blue planetary horizon. The windows that crews look through are set in hulls that withstand drastic exterior temperature swings to protect the station's air-pressurized interior. But all so-called living systems are exposed to innumerable hazards that get in from outside. Radiation shoots into flesh, damaging pieces of what both Earth and space life scientists call cellular systems. Without gravity's pull, blood and other fluids push upward and strain organ systems, while skeletal systems lose bone mass. Nonhuman systems fare the same: disoriented spiders in plastic lab boxes weave chaotic webs, the roots of onboard greenhoused plants grow askew. Despite all this, technicians described the station to me as a systemic platform from which to go into other extremes.

Congressionally authorized during the Reagan administration and flown in the wake of previous U.S. and Soviet space stations, this spaceflight system is a form of national expansionism that is as much environmental as it is geopolitical. How do disparate things become technically associated as "systems"—and to what ends? How has the extreme otherness of outer space played a role in conjoining "systems" and "environments" across spatial scopes and scales? In this book I address those questions by following how spaceflight systems define modern environmental contours and relations—within Earth's bounds and also beyond them. By attending to the practical terms of that work, I shift the frame of a common scholarly analytic focus on spaceflight as a technocratic enterprise, away from rocketry and toward environmental control.

This ethnography is about how U.S. human spaceflight practitioners work with and order "systems" in an environmental cosmos. To examine these processes, I flip anthropology's disciplinary relationship to the system concept. Instead of studying social and cultural configurations *as* systems, I am concerned with how systems are socioculturally configured as situated things. Although the classical Aristotelian definition of a "system" as "a whole greater than sum of its parts" has given the concept a seminal role in modern Western cultural and political history, social scientists do not generally make "system" an ethnographic object per se. To name communication, culture, societies, ideologies, experiments, technologies, and envirotechnical processes as "systems," as a variety of historians and social analysts have, is to elucidate them as modern forms.[1] But even the analytic naming of "systems" intensifies the production of what social theorists Donna Haraway, David Harvey, and J. K. Gibson-Graham collectively refer to as modern "webs of systematicity" through practices that both delineate and exploit connectivity.[2] Although contemporary scholars critique "system" as an organizing feature of Western empire and social control, the concept has not disappeared from social scientific use. Social analysts explicitly attending to the ethnographic dimensions of metarelational posthuman formations such as networks, infrastructures, and assemblages often describe them as being made up of "systems," which further backgrounds the concept.[3] As an omnipresent relational form rooted in early modern philosophy and science, a "system" may be thought of as mechanically lattice-like, organically tissue-like, or even metaphysically amorphous. And it is hard to find an ethnography that does not reference some kind of "system." Like "culture" and "society," "system" is a modern metapragmatic concept; it relates things in spaces that exceed the global and planetary. This is my ethnographic object of study. But rather than producing more systems theory, I take on the problem of how to do fieldwork on and theorize "system" as an epistemic sociotechnical formation—a *relational technology*.

NASA's International Space Station flight control room during the "capture" of the SpaceX commercial spacecraft Dragon. As the first nongovernmental vehicle to dock with the station under the Bush- and Obama-era commercial orbital space transportation programs, Dragon represents the proliferating military, civil, and commercial systems of the low Earth orbit environment. Courtesy of NASA.

## System as a Relational Technology

"System" is—and is more than—an idea or thing; I argue throughout this book that it has also become a technology for putting conceptual and material relations together. Disciplines from biology and engineering to geology and organizational planning deploy "system" to understand things as purposefully conjoined but also to make and retool certain relations among particular known things. So, in calling "system" a relational technology, I take *all* systems as socially made, not given. This ethnography is grounded in the intensive and wide-reaching labor involved in maintaining and intervening in sets of relations called systems. I am concerned here with how such systems make spaces, from bodily interiors to far-off environmental spaces. By attending to how groups work with and control "system" as a relational technology, I join scholars interested in the production of modern technologies of all kinds, such as machinic things and cultured biologies and modes of self-making, that also delineate the kinds of spaces that they occupy.[4]

As relational technologies, systems conceptually and physically formalize social, material, and spatial connections *and* separations, often in terms of a system–environment relationship. The system/environment dyad has become a kind of form/space dyad, but as William Rasch and Cary Wolfe note, it can also be understood as a fundamental modern pairing related to the subject/object dyad. Their observation raises questions about how the coproduced system/environment dyad manifests socially and what it does, especially as it has become universalized within settings in the global North.[5] I follow how elite communities of practice hold "system" and "environment" together to interrelate things in and beyond direct experience. This means I am concerned as much with systems *thinking* as with systems *work.* As the story I present here shifts from work in Earth's watery subsurface to the edges of the solar system, it tracks how governmental agencies and private industries build and link systems to annex outer space as a governable outer environment.[6] Today more systems equal more systematicity—as well as more environmental spaces to know and manage.

In this Introduction, I provide a genealogy of the shift of "system" from relational form to technology of and within environments; I do this also to show how the International Space Station works as an ecosystemic technology of environmental expansion. Space systems such as the station are the centerpieces of NASA's historical sequence of congressionally approved programs attached to administrative eras. This book is about the Constellation program, the interplanetary human exploration program of the George W. Bush era that was slated to go back to the moon and on to Mars and "beyond" between 2006 and 2020, for a projected cost of $230 billion. Although spaceflight policy analysts warned that Constellation was fatally underfunded, NASA leaders strove to elaborate a discursive and material "systems architecture" to coordinate public-sector and industrial sites on Earth and in space.

To fulfill Constellation's "system of systems" architecture plan, workers set out to design spacecraft and comprehensively systematize human bodies, spacesuits, habitats, and nonplanetary objects like asteroids in order to develop environmental knowledge and control. I did fieldwork on these processes from 2005 to 2012 in four NASA centers in Texas, California, and Washington, D.C., as well as in three U.S. and two Canadian medical centers, aerospace companies and conferences, and field testing sites. During this time NASA employed 18,000 civil servant workers and supported the work of an estimated 40,000 contractors. These numbers contribute to making the federal government the nation's largest employer, in part through its strategically underdocumented and reportedly increasing pool of contracted labor. The Johnson Space Center (JSC), while I was there, reported having approximately 3,000 civil servants and 12,000 contractors. Of its civil servant population, most were male (64 percent) and white (74 percent), but as an intern working with non–civil servant contractors, student interns, and spaceflight organizations I also traveled in the more diverse worlds of U.S. spaceflight as an aspirational activity.[7] Systems control in the United States is structured as male and white, but in this book I also call attention to how systems building, maintenance, and reform work becomes differentially socially distributed—that is, socially systematized. In all, I interacted with a total of 121 interlocutors and observed space science, engineering, administrative, and design processes as well as mission planning and training. Although Constellation was canceled by the Obama administration in 2010, its systems architecture groundwork continued into the Trump era. As such, Constellation represents an enduring programmatic shift: from how to fly systems into space to how to manage systemic life across semibounded environments.

By shifting ethnographic attention beyond earthly environmental boundaries, this book examines the changing terms and spaces of systemic sociality as well as politics. In doing so, it demonstrates how, in the United States, "system" is a cosmopolitical tool of empire building and of experimental environmental practice.[8] Since the United States launched its satellite Explorer in 1958 in response to the Soviet Sputnik launch, its spaceflight systems work has encoded space with liberal Western values, such as when U.S. astronaut Neil Armstrong spoke of his first step on the moon in 1969 as "one small step for a man, one giant leap for mankind." But today the "system" concept has altered from its midcentury incarnation as a wholly controllable, closed-off, politically inert thing. It is now also associated with properties of contingently unbounded relational complexity and emergence as well as with politicized knowledge about interdependent life in climatically changing and toxically burdened environmental spaces.

Spaceflight provides a case study of what twenty-first-century U.S. semiclosed/semiopen systems thought and work look like, as a zone of intersecting social and political intentions. Jeff, unlike flight controllers of the past, is a member of Engineers Without Borders, where he and his colleagues imagine the "borderless" portable goodness of systems work within Earth and space environments that are vitally and explicitly interconnected, both cosmologically and politically. NASA, therefore, is a command-control technocratic site, but it is also a social node for people involved in spaceflight both as a job and as a calling and a dream. In this book, I examine spaceflight workers' participation in governmental systems building as well as in speculative practices concerned with systemic possibility and failure. In the United States, "system" as a relational technology is itself an emergent phenomenon; it is linked to the politics of environmental knowledge and control as well as to utopian intentions.

As I offer the following genealogy of "system" as a relational technology, I keep it connected to the kinds of spaces historically related to the concept: the solar system, the globe, environments, and ecological systems. Within this frame, the International Space Station is a system that extends national environments, even though it flies in extraterritorial space that, by international agreement, cannot be made sovereign or owned. This allows me to argue for the value of bringing outer-space-as-environment into political ecological analyses, especially given how systems work remakes the material scopes and perceived scales of ecological relations. I then lay out how the chapters of this book evince that argument and make outer space appear differently as an anthropological field. Since it should be clear by now that I am taking "system" as an object of study rather than as a given descriptor, I ask that you imagine that the word is in quotation marks or italics even when it is not. And that you perceive the terminological repetitiousness and circuit-like connections of my analysis as a thick map of systematicity as a cultural field.[9] To defamiliarize "system" is to attend to how a modern relational cosmos is being produced. My ethnographic approach addresses systemic theorizing *and* practice, closure *and* opening, and tangibility *and* intangibility in relation to the environmental spaces that systems are empowered to fill in and out.

Solar system image that accompanied NASA's 2017 public announcement of a call for solar system–scaled science mission proposals. Courtesy of NASA.

## Relating Things in a Solar System

Back at mission control, I watched a live feed of crew members' weightless working bodies dipping weirdly inside the space station while I listened to simultaneous flight control conversations through a headset borrowed from ROBO—the Robotics Operations Systems officer. He grinned at my reaction to the cacophony. I remembered an earlier exchange I had with Claire, a Habitability and Human Factors Division chief who took part in designing tools for workers in weightlessness. Both ROBO, a young former U.S. Navy service member of northern European and Ojibway ancestry, and Claire, a white middle-aged woman, grew up around the Great Lakes. Claire's NASA contractor father taught her to navigate by the stars and influenced her identification as a "space-crazed adult." NASA centers like JSC are for the most part peopled by civil servants, like ROBO and Claire, and contractors. Many from both groups have experience and family in different arms of government service, showing the role of technocratic generations in reproducing what Akhil Gupta describes as the "bureaucratization of everyday life."[10] After obtaining a degree in human ecology and "bringing three kids on the planet," Claire got an aerospace architecture degree to make "homes with a shell around them." She described her branch's focus under Constellation-era NASA administrator and systems engineer Michael Griffin:

> We looked at how humans interface with the environment from a macro level. . . . There is a general acceptance to say the human is a system . . . as much a system as propulsion, electronics, avionics stuff, like that, but that dehumanizes it a little bit, that's why I tread real light when saying "human system" because there's that "wow" factor that you can't account for.

As the "wow" factor signals, systems work with human beings is not only a form of technical rationalism but also a form of what C. Wright Mills and Veena Das describe, respectively, as "metaphysical" and "magical" modes of technocratic thought.[11] In spaceflight work with interacting mechanical and living systems, the "human system" is, as I describe later in this book, embodied and symbolic, aspirational and problematic; it is a feature of everyday technocratic work that formalizes ways to designate and relate systems in environments.

Scholarly work that elucidates how system became a generic relational form provides the ground on which I describe its contemporary manifestation as a relational technology with a role defining environment as a generalized spatial kind. To understand system as a relational technology, it is crucial to track how the history of system is tied to the invention of spatial scoping technologies, such as the telescope and the microscope—and contemporary things like space stations—that are not just about knowing systems in environments but also about forming them. Following discussion of this history, I introduce NASA as a producer of system–environment relationships at scale and in the extreme. This provides a way to recognize how NASA's power and predicaments as a governmental institution emerge from the conjoined scope of the organization's geopolitical designation as a *space* agency and its everyday systems work as an *environmental* agency.[12]

While sitting at the flight control console, I stared at a schematic of "the human/machine system interface" taped to the desk in front of me. A sketch of a spacesuited figure was centered in an exploded diagram showing the body-to-suit-to-spacecraft interconnections of environmental protection and life support. This was a visual aid for the BME (biomedical engineer) flight controller next to me, who was tasked with troubleshooting relations between astronauts and crew health care systems technologies. The graphic also invoked an iconic NASA-originated system: the "cyborg." In 1960, a psychologist and a musician-inventor coauthored a study of "cyborgs in space," which examined how biomedical technologies could cybernetically adapt an organismic system—in this case a mouse modeling a human—to the extremely "hostile" extraterrestrial environment.[13] As this human–machine hybrid became emblematic of the aims of cybernetics to model and control how systems relate to but stay separate from environments, the cyborg entered popular culture as an inhumanly disembodied or transcendentally embodied being.[14] To make my systems history experience on the console even deeper, the flight controller conversations I was listening to in my headset were called *loops,* a midcentury cybernetics term used to describe system–environment communication "feedback." But, as I detail later in this book, the early cybernetic organism plan was not implemented to modify astronaut bodies. Despite how the things and lingo of cybernetics circulate today, no one I met in NASA referred to astronauts as cyborgs, and most reported little knowledge of cybernetics or formal kinds of systems theory. But *system* and *environment* were everywhere generalized and inextricable terms, staples of conversations, image making, and technical reports. Of the 970,000 documents available on the public online NASA Technical Reports Server as of June 2017, a search on "cybernetics" yielded only 8,000 hits, versus "system" with 290,000.[15] Cyborgic before and after the cyborg, therefore, the system concept has a long history of conditionally connecting heterogeneous things. In this sense, the graphic I saw in mission control was not so much a depiction of a cyborg as it was what John Tresch calls a "cosmogram"—an image representing the elements and shape of a modern Western cosmos: a systemic cosmos.[16]

When I began my fieldwork, "system" was gaining a history as a powerfully universalized and idealized modern form. Recent scholarly analyses of the concept have attempted to make visible its long "hidden history" as a scientific and philosophical metaphor for coherent wholes, a concept that relies as much on its everyday use as a "tradition" of modern practice as on its use as a formal theoretical construct.[17] Caroline Levine argues that modern traditional forms in widespread use, such as "wholes," "rhythms," "hierarchies," and "networks"—and, I would argue, "system"—conjoin with other concepts to shape social experience.[18] When Charles, a civil servant doing International Space Station thermal systems work, said to me, "As a systems engineer I like knowing I can apply all my tools to anything, even human beings," his tooled-together forms of identity and governmentally empowered scope of work pointed to the significance of "system" as a form making all kinds of sociopolitical connections. It is not just a form of "whole" but also a form of *techne,* rhetorical and enacted, which delineates wholeness. Clifford Siskin traces the emergence of the connective system concept to Galileo's telescopic descriptions of worlds as systems, arguing that they established system as a "genre" of worldly coherence and scale that has both shaped and linked modern science and literature.[19] Such cultural historical approaches bring "system" into the analytic foreground as a modern generic form. However, as anthropologist Marilyn Strathern argues, socially shared modern models of interconnected reality rely on the undergirding logic of another generic form: the "relation." Historically as today, "system" and "relation" are interdependent conceptual forms brought together across forms of practice.

Strathern's work on how "the relation" became a "generic" social category in the seventeenth century shows how astronomical theories and technologies connect with modern legal theories and techniques. Strathern describes how the invented social generic of "the relation" in English kinship law formalized a legal "logic of totality" based in principles, forces, and relations that exist "beyond the parts."[20] Strathern, like Siskin, locates this totalizing logic in philosophical elites' aims to appropriate the scientific principles of post-Copernican cosmology, initiated by astronomers like Galileo who worked to reconcile mathematical and telescopically observable problems that required treating cosmic objects as interactive rather than divinely emplaced bodies.[21] Strathern points to how Galileo's work influenced John Locke's philosophical exposition of "infinite" relational logics, describing the epistemic pride of place *relation* now takes as a generic term in disciplines of all kinds, including social science. Having called for social scientists to understand their own embeddedness in modern logics of wholeness as well as in partial connections and separations, Strathern advocates studying how such categorical generics emerge and "behave" in discourse and social action.[22] System is in need of such attention. Strathern's admonitions dovetail with historian Caroline Jones's reminder of how difficult it is for contemporary historians *and* critics of system-as-concept to "turn from" the experience of living within systems-structured modern worlds, where systems "are turning within us, the dynamic engine of our imbrication in many aspects of lived reality."[23]

All of this work on generic forms suggests that along with histories and critiques of the system concept there is room for more examinations of its contemporary social life as a kind of reality-making tool or engine, including ethnographic investigations of how it works to structure relations across domains of ontological and spatial difference. Like all technical things, systems are imbued with cultural ambivalence and put to use for various social ends. The "system" form creates subjects and engenders diverse and unequal modes of association in social spaces. It has a role in making what scholars describe as the modern power-invested relations of large-scale spaces, different worlds, and technoscientific sites that unfold new social and natural spaces.[24] Strathern provides a clue to this when she advises social scientists to "appreciate the membranes of relationality by which the heirs of the scientific revolution assemble, and dis-assemble, their knowledge."[25]

The system concept that gained far-reaching use among elites during the seventeenth century—and that allowed John Locke to coin the term *solar system*—is one such generic thing that was at that time more mechanistic than membranous. From the seventeenth through the nineteenth centuries, influential figures wrote totalizing "systems treatises" spanning the spatial, the physical, and the metaphysical and established systematicity as a technically logical ordering mode. Immanuel Kant's perception of a sublime connection between the "starry heavens" and his own moral and existential consciousness reflects his goal for philosophy to be an ideal-typical system within a post-Copernican cosmology. Historian of modernity Hans Blumenberg describes that post-Copernican system as a "prototypical supersystem" that inspired both Kantian and post-Kantian philosophical theorists to focus on the mechanics of cosmic coherence as well as order.[26] Thus system shifted from astronomy into philosophy without disturbing the existing logics of Western patriarchal theological and racial order. Systematicity became a cultural structuring element and sensibility in works such as René Descartes's methodology; Locke's legal and economic philosophy; Georg Wilhelm Friedrich Hegel's articulation of logic, nature, and spirit; and Alexander von Humboldt's self-proclaimed "systematic" efforts to describe a wholly interrelated cosmos. In this way, systematicity became a hallmark of good thought, authorizing what historians describe as a philosophical demonstration of "completeness" and "order" with the epistemic power to gloss over real-world contradictions inherent in principles like unity, holism, and individuality as well as in binary categories like living/nonliving and conceptual/material.[27]

In the nineteenth and twentieth centuries, system was taken up as an authoritative conceptual form but also became the target of natural and social theorists who observed natural and social systems as less than total or orderly and as subject to internal and external transformation.[28] Using new observational technologies and techniques, such as microscopic study and naturalist fieldwork, nineteenth-century historians, philosophers, and scientists made "system" an object of critical analysis as a concept referring to things in formation and potentially reformable. German philosophers such as Friedrich Nietzsche, Friedrich Wilhelm Joseph Schelling, and later Martin Heidegger rejected the idea of totalistic systems and systemic totalizing. Social theorists including Frederick Douglass, Alexis de Tocqueville, and Karl Marx explicitly critiqued slavery and the economy as systems that produced violent and destabilizing conflicts in social life. As a result, from the late nineteenth through the early twentieth century philosophers and scientists began to theorize systems as purposefully interconnected and tending to equilibrium but also dynamically and spatially indeterminate. This provisionally divested the concept of some of its totalizing closedness and mechanical orderliness.[29] As late nineteenth- and early twentieth-century laboratory and evolutionary biologists took up the system idea, they focused on relations among living parts and things. Biologists began to describe systems as forms in mutually interactive relation also to spaces, in particular to "environments" as sets of surrounding conditions—from households to the planet.

In this epistemic context, the microscope and the telescope became nation-building instruments for producing prestigious know-how about living on a planet in a dynamic rather than quiescent solar system. In the United States, solar system knowledge production had special significance for industrial elites trying to define themselves against an agrarian past. In the mid-nineteenth century, the building of private observatories was associated with antebellum efforts to establish American scientific legitimacy, while later campaigns to relocate astronomical work to universities shifted the locale of cosmological knowledge production from the private to the national sphere. During this time, the solar system gained a scientific and aesthetic presence in elite American geographic and natural imaginaries. Its interacting stars, planets, and powers featured in the nature writing of Henry David Thoreau and the transcendentalist writings of Ralph Waldo Emerson, and also in Victorian-era fascinations with making other planets into alternative worlds with humanoid or alien life.[30] Following mid- to late nineteenth-century discoveries of disorderly solar systemic bodies such as asteroids, early twentieth-century collaborations among geologists, astronomers, and biologists conjoined earthly and solar systemic geological evolution. Coming undone was the notion of an orderly Copernican solar system made up of an empty

Newtonian vacuum predictably ruled by planets, ordered by orbits, resolved of primordial chaos, and within which the Earth stands alone as a removed and unique thing. Later, between 1920 and 1960, "solar system astronomy" and "planetary sciences" emerged to posit a space full of matter in motion that required science and policy to know and manage.[31] Earth was not just a planet *of* environments but was also situated *in* an environment—one that was not sedate, predictable, or separate from Earth.

Emerging within this systems historical timeline, the midcentury NASA hybrid mechanical/living cyborg capable of surviving in any environment represents a hybridization of generic systems concept features: it is a system that is calculably predictable but also has the lively features of an environmentally responsive thing. The key articulation of what would become a generic system–environment relation came from Austrian biologist Ludwig von Bertalanffy, whose "general systems" model of lifelike environmentally interactive things with semiautomatic modes of behavior inspired contemporary cybernetics, informatics, systems engineering, and social systems theory. From this point forward, systems were dynamic vitalistic forms across scales: from microscopic atomic and cellular levels to the macroscopic planetary level.[32] The rise of structuralism and functionalism in early twentieth-century social science represented an attempt to address both the built stability and the contingent lifelike responsiveness of social systems, which were understood to be more or less well adjusted to changing milieus and environments. This conceptualization infuses later social and world systems theories, which are "metabiological" in their assumptions of systemic growth and development.[33] Contemporary feminist, postcolonial, and critical race theorists critique how, as Kwame Anthony Appiah puts it, such developmental organicist logics legitimate the totalizing logics of evolutionism and imperialism that continue to benefit particular groups in a global context.[34] Scholars tracing the influence of the system concept in art, literature, and architecture show how systems thought eventually became problematic because of its association with militarization and "antihumanism" but persisted in concerns about ecological futures.[35] Yet even as the system concept persists as a mode of conceptual capture and ordering, it has also become a way to actively represent life at large as a process of relational emergence. It continues to animate contemporary normative and critical theories of how things hang together and change in science, social science, and philosophy. For example, Alfred North Whitehead's and Gilles Deleuze and Félix Guattari's twentieth-century philosophical theorizations of complex emergence and singularity, which continue to inspire contemporary social theories of relational formations, both rely on the system concept; this signals the concept's continuing meaningfulness in Western thought as an immanent and transcendental relational formation as well as one that organizes social life.[36]

While the late twentieth-century system repetition became a kind of semiopen complex generic form of thought and theory, organizations like NASA, as I show in this book, were working with systems as relational technologies that required steady sources of capital and new forms of organized labor for their production and maintenance. These systems have rhetorical dimensions, like Constellation's plans for permanent interplanetary socialities and economies, and material ones, like networks of communications satellites or the International Space Station. In his history of the 1960s–early 1970s Apollo program, Stephen Johnson describes NASA's systems thought *and* its systems-driven social organization of work as the "secret" that enabled its leadership to control both technical and organizational complexity.[37] During my fieldwork, Mark, a veteran astronaut assigned to a future crew slated to be launched to the International Space Station, explained the agency's "visionary" approach to space systems building as similar to "the mind-set of the interstate highway system." He said this "mind-set" keeps the outcomes of systems work open because "we don't even know, we can't even imagine what we're gonna be about to produce, how we're gonna change [things] for the better." Mark views terrestrial and space-based systems as tools to manage investments in spatial expansion.

Historians have started to focus on system as a technoscientific tool that builds theories of connectivity as well as modes of spatial intervention and control. Darrell Arnold describes twentieth-century "systems theory" as an inherently "constructivist" approach to designing systemic things and processes that had "dramatic effects on various areas of life" in Europe and the United States.[38] Historians of midcentury industrialization and operations research who track human–machine interaction engineering in companies such as Bell Labs show that the notion of a functionally integrated system was promoted as the ideal form for connecting everything in future-focused social spaces.[39] Historians of ecology emphasize how the system concept provided biologists with a way to codify the "assembly rules" of relational holism that both ordered and "entangled" things in an "ecosystem" of systems that still escape full quantification.[40] Utopian visions of cybernetics as an emancipatory tool led to the development of technologies like the Internet and served as the foundation of anti-body transhumanist and immortalist thought, as well as holistic cultural theory and later feminist posthumanism.[41] In scientific and governance systems–focused organizations, the "best" system of any kind is not natural—it is one that can be constructively improved to function better, adapt controllably, scale flexibly, be resilient, remain sustainable, and relate well with other systems and environments.

Mark's labeling of the space-station-in-orbit as a highway-like placeholder for beneficial systems-to-come calls attention to social critiques of spaceflight as speculative rather than productive—and also to systems work as deeply aspirational. NASA's programs are continually subjected to criticism regarding their cost-effectiveness as nationally funded projects that circulate money and benefits most directly to science and industry elites. The station has been justified as a "multipurpose" platform and was recently named a "national laboratory," but it continues to be called into question as a national budget line item across its aspirational systemic purposes: as an excessively expensive mode of global surveillance, international collaboration, and research. Authorization of the station followed hyperbolic promises made in congressional testimony, including that it would lead to a cure for cancer and enable the development of alternative forms of energy in the microgravity environment. Its military–industrial jobs–focused congressional advocates and research networks continue to articulate these justifications despite low numbers of peer-reviewed scientific articles and unclear technology "spin-off" benefits. Scientists and engineers in the United States and Europe who are involved in station research gave me passionate explanations of exploratory discovery as a legitimate mode of work. One biomedical researcher, whose work on a cancer cell–growing bioreactor project was lost with the space shuttle *Columbia* disaster in 2003, told me forcefully in her medical school office, "I wish the public

The Johnson Space Center in Houston, Texas, is, like all NASA space centers, strategically environmentally situated near waterways, airports, and major freeways to receive and distribute systems-building matériel. Courtesy of NASA.

would take a look at the history of what it takes to make the conditions of discovering something really truly new." It is this paradox—that elite systems work is powerful because it can connect things and spaces, but it must be constantly symbolically front-loaded with aspiration—that human spaceflight projects exemplify in the extreme.

Highways and space station systems are differently productive, but both enact ideas of spatial control; this book highlights the coconstructively material and symbolic power of systems work. The International Space Station flies as a flagship of constructed and constructive systems, visible in its role in defining the management of two kinds of idealized late industrial spaces: the "total" environment and the "extreme" environment.

## The Total and Extreme U.S. Environment

During my fieldwork in Florida on a spaceflight training program (the subject of chapter 1), Beth, a white female aerospace psychologist long involved in crew support and biomedical research for the International Space Station and the space shuttle program, taught me the dimensions of space's environmental extremity. She raised a closed fist and then unfolded her fingers one by one: "threatening, isolated, confined, remote." Beth's enumerating gesture hails the kinds of spaces, persons, and even systems involved in the contemporary U.S. "extreme" category: prisons, disasters, "deviant" social groups, forced and voluntary environmental experimentation on people and animals, military and nuclear risk management, polar spaces, underwater sites, and space exploration. Beth had also worked in the Federal Aviation Administration, and her governmental career can be seen as a node in the supraglobal social and political life that powered systems enable. Sheila Jasanoff argues, in her history of science as a social ordering activity, that "powered flight" is one among the "salient" technoscientific ordering "powers" that could orient a global political history of the twentieth century.[42] As a result of technologies that fly human bodies and senses into inhospitable spaces on and off the planet, outer space is not considered a unique kind of environment in the U.S. today—it is one among a set of environments categorized as *extreme.* Here I introduce how spaceflight works with extremity in terms of system–environment relations, but also as a charismatic, edgy dimension of a larger U.S. cultural effect. U.S. spaceflight's production of alluring extremity is a motif in this book. It allows me to highlight a cultural shift in which "normal" has become associated with normative health as well as with developmental stagnation, and "extreme" has become associated with threats as well as with "improved" social and ecological capacities.

The now-common treatment of "environment" as a concept that can be divided into categories, such as normal and extreme, was bolstered by mid-twentieth-century work aimed at knowing and governing "the" total environment—work enabled in part by government-sponsored human spaceflight.

## *Space as Total Environment*

In NASA language, the flight controllers and crew members I watched during my time in mission control were engaged in "station tending," the daily activities necessary to keep things running and to remake the boundary of a postterrestrial "total" environment. Like the orbiting station, the first human-made satellite, Sputnik, was not reducible to simply a military gesture. It was also a total planetary surveillance machine that debuted as part of the 1957 International Geophysical Year. In the United States this event turned the National Advisory Committee for Aeronautics into an aeronautics *and* space administration. Legal experts designed the new agency to be apart from the Department of Defense's space projects. As "pioneer" space law expert Eileen Galloway has recounted, "We needed a total space program" with a separate civil agency that could develop the "many benefits there were to space" outside the parameters of military law.[43] Five years after Sputnik, American biologist Rachel Carson warned Congress and the public that "universal" forms of modern industrial contamination were threatening "man's total environment," including that generated by aerospace activities.[44] A year after Carson's warning, philosopher Hannah Arendt predicted that a wholesale "conquest of space" was likely to be stymied by space's effects on human bodies, but she cautioned that spaceflight's real danger lay in its alignment with other technological efforts to "dispos[e] of terrestrial nature from the outside."[45] Arendt predicted that artificial energy-making and life-manipulation processes would decenter Earth and "man" within a new technically "elemental" universe devoid of humanistic meaning and ethos.[46]

Taken together, Carson's and Arendt's warnings point to aerospace's paradoxical role in both defining and contaminating total environments as well as in situating a generic "human" within systems that both center and decenter it. One of these environment-defining processes occurred as the nuclear and aerospace industries produced data about the circulation of bomb radiation and other forces that validated ecologists' claims about ecosystemic connectedness.[47] Outer space became the occupied exterior environment of "the environment" of public discourse, as well as a futuristic systems development space for Cold War security and commerce. Today, significant investments of tax dollars and market capital continue to define the characteristics of a total geospatial environment. In 2014, the Organisation for Economic Co-operation and Development released a report comparing national investments in civilian and military aerospace spending, noting that the United States stands out as the most invested, with its combined space-sector expenditures of $40 billion per year.[48] During my fieldwork, NASA's funding fell from 0.7 percent of the federal budget to 0.5 percent, which still amounted to $1.9 billion in 2016.[49] Since I started this project, Japan, India, and China have shifted the space superpower field with expanding remote sensing and human flight programs. As a result, the provinces of space and environmental policy are converging as Earth orbit fills with an increasing load of productive machines and hazardous debris.

With participation in the International Space Station mandated as a Reagan-era exercise in improving international relations and creating a long-term spaceflight environment, the station's American-made parts were produced by companies that had started the Cold War space age using imported German government engineers after world War II. Although NASA is tasked with making future spaces, it has also invested in historical documentation and memorialization of its projects. Members of a productive network of NASA and non-NASA historians and policy analysts produce much of the scholarship on the American "space age" from midcentury to the present. They explain its trajectories according to the influence of three general factors: the relationship between political leaders and NASA administrators involved in the agency's "reinvention," an increase in space bureaucracy along with technical growth and decline, and the relationship between policy making and technology building in changing sociocultural contexts.[50] Scholarship concerned with the last of these factors mounted as U.S. spaceflight "fiftieth anniversary" milestones began to be marked at the beginning of the twenty-first century, sponsored in part by NASA, whose history program put out calls for multidisciplinary assessments of spaceflight's "societal impact."[51] These calls came after decades of retrospectives published by popular and academic presses, including biographies, program histories, and cultural critiques of the space age as a "dream" and a "mentality" as well as a visionary or failed historical moment.[52] While human spaceflight projects move forward in fits and starts over time, orbital and robotic space power continues to expand the spatial reach of national spaceflight systems.

Space superpowers amass prestige and capability through expensive, labor-intensive projects like orbital space stations that, among other activities, push the boundaries of national environmental management. These projects began in the early 1970s, when space race goals shifted from landing heroic subjects on the moon to running low Earth orbit as a globalized space. In the 1970s, the United States and the Soviet Union started provisioning inhabited stations for protomilitary occupation and technological development. Stations like the Soviet Salyut 1 and U.S. Skylab were sovereign machines, but later stations that arose from international collaborations made space a site for diplomacy and technical systems harmonization. As a product of decades of social interaction beyond national borders, the International Space Station exemplifies the "modularity" that Hannah Appel observes in offshore resource-extraction structures and is an environmental extension of what Keller Easterling describes as the "overlapping or nested" forms of sovereignty that create the spatialized systems and infrastructures of "extrastatecraft."[53] Since 2000, the station's twenty-six member nations have been sending crews to inhabit it continuously. It is valued at US$100 billion, costing millions per year and taking thousands of people to maintain. The station and the space shuttles that ferried equipment and crews into space until 2011 have been dubbed the most complex machines ever built, in part because they are encapsulated life-support environments. They are, as one middle-aged white male shuttle astronaut and physician told me in a low, reverent voice, "the most beautiful systems ever built." Despite debates about their cost and value, space stations continue to orbit as attractive arms of national environmental control and as bearers of collective dreams for unbounding built national systems.

In the late 1990s, Russia prepared to launch the International Space Station's first habitat module into orbit to make a supraglobal transnational living space. Writing during that decade, Denis Cosgrove argued that the production of a modern global geography had become "intimately connected" to outer spatial cosmography, and political historian Walter McDougall narrated how the space race had scaled the global "spheres" of state-sponsored competitive "command technology" research and development.[54] Today scholarly attention to how geographic verticality, volume, and scale are being produced at embodied and sociocultural levels provides new windows for analyzing categories like "global" as, in the words of Bruce Braun, "an effect not a condition" of various space-defining processes.[55] While mid-twentieth-century designations of outer space as a speculative "frontier" index colonial and manifest destiny rhetoric, they are also the effects of technical processes mediating space and Earth linkages through machines as well as through bodies. Patricia Limerick describes the authoritative imperialist as well as popular rhetorical meanings of social boundary reforming packed into the "frontier" metaphor that became attached to space in the United States, and Peter Redfield's ethnography of a colonial prison and European rocket launch site in French Guyana reveals structural relationships between the technical control of "frontier" bodies and outlying spaces.[56] As Lisa Messeri writes in her ethnography of exoplanetary discovery, the space frontier has become an "outer place" full of other scientifically comparable worlds.[57] But by this time space has also become an "extreme" environment.

## *Space as Extreme Environment*

The "extreme environment" first appeared as a category in nineteenth-century evolutionary biology, and it became a biomedical category in the 1960s with psychological evaluations of military and scientific "performance" in polar and aerospace extremes. But in 1976, NASA's publicly released "bicentennial report" brought environmental extremity into the here and now:

> The population and technological explosions that have occurred so drastically since the nation began 200 years ago have modified your environment in the extreme. . . .
>
> Of themselves, these changes are neither bad nor good. It is what we make of them, how successful we are in adapting our environment to our goals—and in adapting ourselves to our environment—that we create better or worse conditions for human life. The realization that our environment is threatened, that our natural resources are finite, need not be cause for despair. Conversely, they teach us an appreciation of human ecology, the interdependence between man and his surroundings.
>
> Spinoffs from space technology are helping to improve that ecology.[58]

This statement invokes the promissory rhetoric of systems building, part of the history of the rise of systems thinking as an industrial failure remediation process (as I discuss in chapter 1). But it specifically relates spaceflight systems development to the urgency for governmental management of increasingly extreme everyday conditions. In a sociological analysis of extreme environments as modern sites of social management and change, Kroll-Smith, Couch, and Marshall argue that contemporary researchers and governments have become equally concerned with producing knowledge about spaces that "elud[e] efforts at normalization" and that expose not only the vulnerabilities of bodies and spaces but also "aspects of social systems and processes that are typically obscured by the routinization of everyday life."[59] Today, the extreme environment category is invoked across domains of discourse and practice and makes government-sponsored spaceflight a systemically and environmentally developmental enterprise.

While social scientists and postcolonial scholars have described outer space's role in effecting the spherical "earthly politics" and "satellite planetarity" of a global environment, the processes that effect space's shift from a disconnected unearthly zone to a charismatically connected "extreme environment" have been less examined.[60] Environmental politics historian Timothy Luke traces the first use of the holistic term *environment* in the popular media to a newspaper article on astronautics, but spaceflight historian Roger Launius has also noted that the history of astronautics' particular environment—the "extreme environment"—remains underanalyzed even in space studies.[61] It is outer space's contemporary designation as an extreme environment that creates and shapes space sciences' distinctive concerns with modern life's limits and potentials. And while "extreme environment" is a development-focused category like globe and frontier, it shifts the focus of expertise from territorial to ecological technologies. Claims that spaceflight's value inheres its capacity to "spin off" its technologies, as in the NASA publication quoted above, are grounded in evolutionist understandings of environmental "extremity" as a stimulant to progress. Such aims situate spaceflight within historical and futuristic valuations of the extreme.

U.S. spaceflight and its work connecting built and natural systems in extreme environments contribute to configuring "the extreme" as a cultural concept that celebrates the capacities to control deviance and outlier spaces and to produce social distinctiveness. Discourses about social "extremism," such as terrorism, confer normality to Western imperialism and cultural hegemony. At the same time, the cultural concept of extremity has entered contemporary discourse to signify a condition Edward Said identifies as "something beyond the everyday . . . whose core is precisely what can be experienced only under relatively severe and unyielding conditions."[62] Today, extremity motivates political and ethical activities, as well as forms of commodification and millennialism. People get "into" extreme sports, extreme foods, and extreme makeovers. Western space-age and environmentalist visions also colocate in concerns about extremity. Contemporary environmental activists imagine a return to an original Earth of stable norms and extremes, such as NASA climate scientist James Lovelock's "Gaia," or an arrival at an utterly extreme one, such as Peter Ward's vengeful "Medea" or Bill McKibben's postapocalyptic "Eaarth."[63] Plans to control anthropogenic environmental extremity are invested with hope but also with fears of what Anthropocene-oriented geologists call "a different trajectory for the Earth system" in which Earth becomes as uninhabitable as other planets.[64] Extremity, therefore, is not only threatening but also generative.

The late nineteenth-century biological research that established the relationship between living systems and environments defined "extreme environments" as spaces that prohibit the flourishing of life but can also select organismal capacities for survival and adaptation. In his ethnographic studies of marine and outer space microbiology, Stefan Helmreich describes how discoveries of organisms thriving across a spectrum of statistically normal and extreme conditions reorient biologists from the study of life-forms and norms to the study of "forms of life." He also observes how astrobiologists looking for alien life cultivate a sense of "extraterrestrial relativism" in which biology is universal as a category with varieties of normative forms rather than deviations.[65] Historians of biology and medicine concerned with the production of normality as a scientific statistical category describe how medical and institution-building authorities have attempted to engineer social environments to make normality and health equivalent.[66] At the same time, national practices of exploratory environmental annexation, from the poles to space, began to make "abnormal" environments into governable sites.[67] Although the "extreme environment" category favors an anthropocentric notion of what counts as normal conditions, astronauts and spaceflight activists I spoke with referred to themselves as "extremophiles," having become captivated by biology's term for innovative adaptive proclivity. I found a love of environmental extremity to be widespread among spaceflight experts, who might also have topophilic attachments to particular places, like Mars or the moon.[68]

In spaceflight, extremes are both dangerous and desirable. In the station flight control room, I listened to flight controllers as they managed conditions they defined as extreme for things in their systems. Flaring solar radiation was extreme for communications as well as for bodily systems. However, all the station-bound and returned crew members I met had trained in simulated and field-based extremes. One astronaut I got to know during her NASA Extreme Environment Mission Operations training in an underwater habitat in the Florida Keys became the "women's world record holder" for continuous underwater living and later the "first mother" on the station in 2012. Like other U.S. astronauts marked by race or gender as "firsts," Nicole Stott is often described as a frontier maker. But she also participates in a water-system-engineering nongovernmental organization (NGO) called Geeks Without Frontiers, an organization modeled on Engineers Without Borders that displays the social movement characteristics of U.S. spaceflight advocacy.[69] She has publicly described space as an environment in which spaceflight "shifts" attention from "familiar things" of "Earth in 1 g" (e.g., normal gravity) to "how cool brains and bodies are, how they adapt to being in different environments to overcome challenges."[70] The astronaut's body conjoins frontier and environmental spaces as sites of ordering and control. Other astronauts I spoke with emphasized extreme environments as "technology accelerators" as well as catalysts for human evolution. This imaginary of evolutionary outer spaces makes astronauts kin with a humanoid cultural entity Debbora Battaglia calls the "de-exoticized alien"—an extraterrestrial figure that has become a familiar member and trope of the cosmos-inclined U.S. entertainment and spiritual worlds.[71]

If frontier-making subjects such as pioneers embody settler colonial rhetorics of progress, beings associated with extreme environments—such as deep-sea microorganisms, astronauts, and aliens—have come to embody scientific and social fascinations with the biology of developmental capacity.[72] As spaceflight systems put machines and bodies, human and nonhuman, into space, they remake national ecosystemic boundaries. As a result, the extreme environment of space is not a spatial limit but a political horizon.

## The Political Ecology and Scalarity of a Solar Ecosystem

A year into the start of the Constellation program, I received e-mails from spaceflight engineers and life scientists alike alerting me that President Bush's science adviser, physicist John Marburger, had declared that presidential space policy affirmed a decision to "incorporate the Solar System into our economic sphere."[73] This phrase tied together the rationales for U.S. environmental and economic solar systemic extensions, which found expression in the Bush and Obama administrations' support for new exploratory destinations and commercial space companies that relate the West's two "eco" systems—ecology and economy.[74] In this book I show how the Constellation program's boundary-crossing technologies and engagements with new targets like asteroids turned space into a site of socially uneven environmental access and speculation. In doing so, I follow contemporary political ecology's concerns with tracking changes in what count as environmental sites and ecological objects. To date, outer space has counted as a U.S. political geographic site when scholars examine how spaceflight centers were strategically situated to advance "backward" Southern and Western landscapes and to integrate the government's technical expert labor force, or how spaceflight infrastructures of launch sites and tracking systems connect imperial projects across the globe. However, compared with military and nuclear industrial spaces, U.S. spaceflight's spaces on and off Earth remain much less likely to be investigated as political ecologies. Including spaceflight in this frame involves considering its sites and things to be dually ecologically and economically systemic.

As outer space became an extreme environment in the twentieth century, its inhabited spacecraft and spacesuits became something other than vehicles and clothing: they became kinds of semiclosed ecological systems. Although ecology is not a formally institutionalized spaceflight discipline in the United States, it shares with spaceflight a focus on the scope of organism–environment relations. As historians of ecology and environmental science have shown, the "ecological system" and "ecosystem" ideas are boundary concepts invented by early twentieth-century biologists and shared within the postwar funding and knowledge networks of ecology, nuclear science, cybernetics, environmental design, and aerospace. "Big ecology" and ecosystem science emerged in these networks, as scientists and engineers collaborated in controversial government-funded projects to model biological relations on a global scale. Big systems modeling of what Frank Golley describes as the "loose" and "open" domain of ecosystemic space gained adherents and detractors.[75] As art historian James Nisbet argues, conceptualizations of systemic openness consolidated scientific and artistic interest in the systemic interrelatedness of life, nonlife, and energy as equivalently "vital" components of functioning wholes.[76] These wholes scientifically include outer space, but environmental social science and political ecology have not tended to be as analytically inclusive.

Although Earth's atmosphere is often assumed to be the de facto boundary for environmental politics, outer space science has been defining the scope and scale of those politics. NASA Venus–Earth comparisons and Earth satellite data founded climate change research and politics in the United States, but social scientists still tend to bypass analyses of space's role as a political environment in favor of focusing on earthly politics and terrestrial climatic data regimes.[77] Despite political ecological work to understand how, as Arturo Escobar argues, nature is connected beyond essentialist prescriptions of its boundaries, outer-space-as-nature usually ends up implicitly *dis*connected from debates about nature-culture and multinaturalisms.[78] As Nigel Clark writes, modern liberal thought continues to avoid the problem of the "radical asymmetry" between the cosmic scope of "inhuman" nature and the social.[79] As a result, space tends to stand outside social scientific analysis as a residual nonenvironmental space, or what Erik Swyngedouw terms a "depoliticized environment" of policy rather than everyday politics.[80] In addition, when historians have addressed NASA's concern with environmental politics, they have done so on the level of institutional strategy.[81] Yet I found evidence of diverse forms of environmental perceptions and politics among my spaceflight interlocutors. Anthropologist David Valentine has as well, showing how "New Space" entrepreneurs, differentiating themselves from the "Old Space"–industrial complex, are using capital to understand how to engineer off-world self-contained capsule environments, a process that politicizes nongovernmental environmental knowledge perception and production.[82] Peder Anker locates the political history of space capsule building as a site for technical forms of earthly and nonearthly "ecological colonization" in the beginnings of spaceflight experimentation.[83] Spaceflight's ongoing politically motivated production of ecosystems in extremis opens a window on the ever-shifting edges of environmental power and on the political plurality of environmentalisms occurring in Earth work spaces and in what Debbora Battaglia calls the uncanny conditions of "space itself."[84]

Lack of critical social scientific attention to outer space as environment is also an effect of what political ecology scholars Anthony Bebbington and Jeffrey Bury identify as a "surface bias" that does not adequately account for the postterrestrial scaling of social ecological and economic space.[85] Related critiques call for changes in basic units of political ecological analysis, suggesting shifts of attention from households to bodies, from economic activities to affective economies, from national to transboundary spaces.[86] Spaceflight involves all of these in attempts to scale systems from the micro to the macro level. Even if one obvious macro-scale effect of spaceflight is the securitization and capitalization of outer space, most social analyses of violent and economized environments leave it outside the frame of analysis. Yet, space today is known as an environment that links Earth and space weather, as a system of forces, and low Earth orbit is considered polluted by orbital technology debris. And space is changing as an economic site. New U.S. laws enabling space mining and energy extraction are changing how space sites and objects, from Earth orbit to near-Earth asteroids, are understood economically. Such changes, which I address in this ethnography, compound the global asymmetries of space power and also make them postglobal and extensively extraterrestrial as well as extraecological.

The International Space Station is a large-scale space ecosystemic object, but is also held up as an example of a global "platform" for solar systemic habitation, despite its exclusive membership. Such politically scalar things are increasingly critical topics for social scientists, even if space and its object remain out of the picture. Political ecologists and anthropologists have exposed Western productions of scale as a controlled ordering and nesting of social and ecological life that justifies differential distributions of power and resources. These critiques focus on scale as a mode for making graduated differences in sizes and forms, based in part, as anthropologist Anna Tsing points out, on the biologically erroneous technocratic conceit that living things are homogeneously "scalable," as designed objects are.[87] Richard Howitt points out that while scale may be understood this way, it also signifies a quality of extended relations beyond particulars.[88] It is in this spirt that Clifford Siskin describes the system genre as a "scalable technology," meaning that it acts as a "semantic platform" for making graduated and nested scales of parts and wholes ad infinitum.[89]

While spaceflight technical experts do deploy scale and scalability as precision tools, I detail in this book how they also build equally powerful sensibilities of *scalarity.* What I mean by this is that spaceflight practitioners and advocates routinely make things and processes seem contiguous at scale, such as linking home spaces with planetary spaces, human consciousness with cosmic systems, and human footsteps on planets with human evolution. Rather than precisely continuous arrangements of scalar relations, these affective and speculative jumps can represent what Timothy Clark describes as a scalar "derangement" of political, scientific, and experiential elements that nonetheless becomes perceivable and legible.[90] Thus, in the chapters that follow, I attend to how spaceflight systems work supports imaginations of environmental relations and scalarity beyond everyday experience and also produces the extensive politics of systematicity.

## Political Systematicity

Like other modes of relationality, systematicity takes different forms, of which human spaceflight is one. The chapters of this book are sited where people work with and within systems where "the human" is situated as a distributable working element. In that work, outer space is not just a place to make earthly or to bring down to Earth, but a space in which unearthly differences shape more-than-earthly system building. The system concept provides scalarity for the "relational imagination," creating not only a spatial but also an explicitly politicized environmental systematicity that extends ever beyond.[91]

My field sites revealed how social groups were maneuvering within a changing "civil" outer space policy environment full of new natural and technical things to manage. This book, with its depictions of civil/military boundary gradations that I experienced with other spaceflight workers, represents "civil" spaceflight not as something virtually unreal with a real military underpinning but instead as an actual social world.[92] A small number of my interlocutors were former members of the military, but most were not. Many had the same familial, economic, or structural relations to the military that connect many U.S. residents to the quotidian militarization of U.S. social life. NASA's civil social world was, as I heard often, not just big science but "an organization by and for engineers." During my fieldwork, NASA's engineers and scientists were younger on average than their counterparts in other aerospace workplaces but aging relative to the rest of U.S. work sites and part of a shrinking employment sector. In 2010, two-thirds of the male civil servants at JSC were classified as engineers or scientists, as opposed to about half of the women, of which the rest were administrative and other kinds of technical staff—groups I also interviewed as integral to the production of the civil spaceflight world. Men of all reported "race/ethnicity" groups were equally likely to belong to either the engineer or scientist professional category.[93] Young people I interviewed at JSC represented a continuance (although slowing) of NASA's historical engineering and science intern programs. Managers and administrative staff I interviewed told me that such programs used to be pathways to civil service, but they are now gateways out to the private sector. I also gained access to NASA as an intern, if much older than average.

I recruited interlocutors using time-honored nonsystematic ethnographic methods. Because NASA projects span disciplines and spaces, I connected with interlocutors nationwide through situated chains of contact and invitation. This flexibility was important as I moved through institutions requiring security clearance and other tokens of access. My approach facilitated informal criterion and purposive sampling strategies that allowed me to approach potential participants based on professional and demographic characteristics, so my JSC interlocutor pool was comparable to the center's civil servant demographic profile.[94] One of the key venues I used for networking was the weekly informal "Exploration Faithful" meet-ups created by a core JSC group in the 1990s to sustain human spaceflight institutional memory. Although most interlocutors consented to my use of their actual names, I refer to many here by pseudonyms (single names) while using the real (full) names of others; as I was not allowed to collect demographic data, I either provide or omit information on gender and race/ethnicity based on informed or relative (possibly erroneous) certainty.[95] I sometimes obscure locations and events even though I was not required to do so.

The first chapter, "Metasystem: Making Spaceflight's Parts," is about how spaceflight produces environmental systematicity. Based on my time working on an underwater "environmental analog" for spaceflight testing and training, I examine particular modes of thought and practice that support systems thinking: analogizing, processual temporality, and the making of systemic parts and social participation. These modes socially link systemic things to other systems, like education, environmental and ecological knowledge, national technological infrastructures, national character, and imagined solar systemic scaled futures. Chapter 2, "Connection: The Ecobiopolitics of Space Biomedicine," is concerned with how astronaut lives and deaths are made ultrasystemic even if the astronaut is not a fully knowable and predictable "human system." I show how, in an inversion of the common modern biomedical model, astronaut subjecthood is fundamentally ecological rather than biological. This makes visible what I define as the ecobiopolitics of space biomedicine, in which investments of power and knowledge shift from life itself to processes of ecologically systemic interrelation where life and environment are dually problematized. The third chapter, "Separation: Building Partial Boundaries," calls attention to how systems building is more than the designation of connective wholes; it also requires the production of separation technologies that make desirable forms of disconnection, including the two I analyze together: new spacesuit designs and aerospace environmental cleanup programs. Chapter 4, "Transhabitation: Universalizing Architecture," examines how activities to make space habitable create transitive and developmental notions of home and nation as ecologically systemic spaces. The final chapter is "The Solar Ecosystem: Postterrestrial Ecologies and Economies," which examines the historical and political role of nonplanetary matter—asteroids and comets—in the making of discourses about an ecologically epochal planet Earth and an explorable and exploitable cosmos. I describe how near-Earth asteroids and comets are now astronomical as well as environmental objects—threats but also material resources within an emerging heliospheric ecology and economy. Through these chapters I move across a contemporary transterrestrial space of pasts, presents, and futures.

In the United States today, space and Earth are being strategically systemically connected and separated in scientific, technical, and political terms. Through these processes, "environment" takes on a shifting spatial and imaginative scalarity—more than terrestrial but not utterly unearthly and more than social but less than totally inhuman. In such schemas, "system" works as a quotidian and spectacular relational technology of reality building. I portray these processes anthropologically by assembling the histories, social encounters, texts and discourses, and governmental public images involved in producing a cosmos of systemic environments. This book, then, is an ethnographic cosmogram—a single-handedly written but deeply socially interdependent representation of a contemporary space.

# 1

# METASYSTEM
## MAKING SPACEFLIGHT'S PARTS

We are on our way to greet a crew of six people last seen on earth eleven days ago. This crew has been living underwater to participate in a spaceflight "analog" mission off the coast of Florida. I am on a boat loaded with scientists, engineers, and technicians motoring out to sea to retrieve the crew after they swim up from their stay in a seafloor "lunar base." These analog missions make Earth like outer space, so that earthbound people and things respond as if in extraterrestrial conditions.[1] I am a NASA intern working with mission control staff and participating researchers in a dockside Key Largo marine research center. This makes me one of among many people on and off the planet who are understood to be part, as a spaceflight advocacy story goes, of an evolving exploratory species. As our boat passes through the Key Largo harbor channel to make a four-and-a-half-kilometer ocean trip out to the mission site, long, striated clouds extend into a rolling immensity of gray-white seawater. Out there, twenty meters below the surface, a bus-sized submersible laboratory belonging to the National Undersea Research Center lies bolted onto a reef in a national marine reserve. The Aquarius habitat is rented out for academic, military, and privately funded missions—one terrestrial node, like the International Space Station, in a space systems network.

The NASA program that rents the Aquarius habitat several times a year is known by the acronym NEEMO, for NASA Extreme

Environment Mission Operations. For years, the coral-encrusted Aquarius has stood in for a spacecraft, and its warm, lively, terrestrial sea environs morph analogically into cold, dry, lifeless outer space.

When I began my fieldwork, "analog systems" were important developmental programmatic parts of U.S. spaceflight systems, as described in the report *NASA's Exploration Systems Architecture Study.* This 2005 document outlines a system-of-systems to send humans to the "Moon, Mars, and beyond" at a cost of "more than $2 billion per year in exploration systems, human, and nuclear-related technologies" over a span of twenty years.[2] The report authorizes the continuation of space analog systems and other Earth-based alliance-building projects that have been part of space programs since their inception. The analog mission ending today, the twelfth NEEMO and the second of the three I worked on in 2006 and 2007, is a part of NASA's ongoing scheduled overlap of flight-related work and public outreach on and off Earth. According to the mission schedule, the NEEMO 12 crew was "timelined" to "return to Earth" with an ocean "splashup," as opposed to the "splashdown" of twentieth-century space capsules. As our boat slows near its destination, it crosses a ring of colorful "no trespassing" floaters. In the middle of them, bobbing atop torpid waves, is the top end of a machine understood to function as Aquarius's "lungs." The tall, weathered yellow buoy with "LSB" (life-support buoy) painted on its side emits a generator-powered rumble that competes with the rushing sea wind. The buoy's compressors pump humid surface air down into the habitat/spacecraft via a hose referred to as an "umbilical" line. This Earth-based space program is structured by such intensive analogizing. But NEEMO has broader relation-making powers that extend its analogical reach far beyond this particular location off the U.S. Atlantic coast.

The NEEMO 12 crew's return to the surface marks the end of their time as corporeal astronaut-like human analogs dependent on an extended set of mechanical and social systems. From day one in the habitat, two and a half atmospheres of nonterrestrial ambient pressure began saturating their tissues with gases. This transformed them into "aquanauts," in the space-age lexicon of industrial and military diving. Although it is a short scuba-assisted swim up from the habitat to the ocean surface, the crew members are technically a long time away from Earth. It takes eighteen hours for aquanaut bodies to decompress in

Japan Aerospace Exploration Agency astronaut/aquanaut Satoshi Furukawa stands in weighted boots outside the Aquarius habitat in a public affairs communications image for NEEMO 13. Courtesy of NASA.

the habitat's pressurizable interior compartments before they can safely ascend. As NEEMO managers and participant veterans told me over and over, without such intensively managed systems, "you die."

The NEEMO analog mission program demonstrates two earthly NASA activities saturated with the politics of national and transnational space making and futurity: incorporations of outlying "extreme" spaces and public productions of systematicity. Systematicity, as I described in the Introduction, is a modern ordering schema that relates different parts and makes those interrelations sensible as systems both of and with other systems. In this chapter I track three particular cultural processes that make systems of systems and systematicity spatial and social: the making of systemic *analogies* that align forms and functions across spaces, the production of developmental *processual time,* and the ordering of environmental and social *parts and participation.* These processes are characteristic of modern systems thinking and work, but they also connect systems work with other modern social practices. As a result, such metasystemic processes amplify the meaningful presence of NASA's astronautical systems in relation to other systems, expanding the interrelated boundaries and objects of environmental systems thinking and practice. Therefore, my concern here goes beyond the making of particular technoscientific or envirotechnical systems or the consolidations of infrastructure. My aim is to elucidate understandings of U.S. dual-use governmental systems that support the making of broader systemic formations. Aquarius is a multiuse system made to operate within broader system-infused processes, including military testing, ecological research, environmental science and technology, biomedicine, adventure media, and education.

By attending to NEEMO's metasystemic dimensions, I intend to shift critical analytic attention from systems in and of themselves to the ways that systematicity is made perceptible and distributed as a vital property of national environmental power.

## Metasystemic Processes

In everyday systems practice, the *meta* in the term *metasystem* refers to a "system of systems," but it also dovetails with the ways the prefix is used in social and cultural theory. In both cases, *meta* points to overarching and undergirding dimensions of any particular cultural category. In systems engineering and science, *metasystem* can refer to a system that coordinates or "governs" other systems, where *govern* (with its root in the Greek *cyber*) is an etymologically definitive feature of cybernetic theory that signals how system–environment interactions are organized by regulatory modes of communication and functional cohesion.[3] More rarely, but in this spirit, engineers use *metasystem* to refer to the broader sociocultural systemic processes that define and contain technical systems.

In this chapter I focus on the analogical, processual, and participatory dimensions of spaceflight system making that connect spaceflight practices to broader processes and spaces in order to get *beyond* the more common social scientific idea that systems are defined primarily by closed, internalized cultural logics. I draw attention to how such cultural processes associated with—but not limited to—systems practices legitimate the social coherence of systematicity as a governing relational mode. In doing so, I follow other anthropologists using *meta* to illuminate the broadly arrayed relational dimensions of cultural processes.

For instance, Sherry Ortner's "metasystem" concept designates overarching notions of cosmological reality that make everyday order sensible; in NEEMO missions, *analogizing* situates earthly systems within a scalar cosmological reality.[4] Along these same lines, Greg Urban's system concept–informed "metacultural plane" is a virtual zone where the ideologically and economically exploitable cultural properties of new cultural objects, such as new movies that reassemble parts from extant cultural objects, enable those new objects to circulate as parts of an overarching cultural whole beyond everyday perception. Urban describes this process as when "something that is the property of a thing in the world appears to be a property of a system that is thingless."[5] Similarly, the NEEMO analog publicly reinforces, beyond itself as a systemic thing, systematicity as a spatially and temporally transcendental *processual* condition. Importantly, the metasystemic and metacultural require not only things connecting across social dimensions and scales but also people and signs.[6] Thus, NEEMO's enrollment of *participants,* from researchers to publics, matters as a meta aspect of its systematicity-building methodology. George Marcus, James Faubion, and Dominic Boyer's concept of "metamethod" points to ways researchers communicate a conceptually researchable world; in this case, NEEMO and its participants communicate the existence of an empirically systemic environmental cosmos.[7]

NEEMO lends itself to an investigation of the metasystemic because space analogs are both celebrated and critiqued as platforms that exist to make perceivable connections between the Earth and the rest of the solar system rather than to produce new scientific or technological discoveries. As Lisa Messeri shows in her study of planetary scientists enacting Mars analog missions on Earth to make extraterrestrial space more place-like, such work is as much about producing a socially "generative medium" that creates place-to-place correspondences as it is about practicing science.[8]

The metasystemic processes I describe here are not the only features of environmental systematicity—but they are essential ones for a couple of key reasons. First, they interact to provide shape to otherwise amorphous systems. While in this chapter I have to attend to them in a linear way, which is a constraint of Western writing/reading structures, these processes are concurrent and recursive practice processes that fill out people's senses of systemic space and time and of their position within systems. Second, these processes give systemic relations culturally perceived material coherence and scalarity. For example, analogizing environments and systems in terms of form and function is a tool of spaceflight design and research. At the same time, astronautical authorities also make rhetorical analogies to justify spaceflight's nation-building purposes by comparing its promising forms and functions to historical and current everyday things. NEEMO, therefore, as an analogical system of temporally serialized missions that aim to demonstrate spaceflight's evolving part in modern social life, produces knowledge about how and why to expand national space and manage it as a futuristically inclined environment full of related parts. Based underwater in a charismatically alien but familiar space, it showcases the extreme scalarity of what NASA's Human Research Program calls "human–system combinations": humans living in or working with machines in environmental contexts.[9]

At the time of this writing, NEEMO missions are still being conducted, and the program's developers continue to tout the Aquarius habitat as the most environmentally analogous analog of all. Being made a part of all this as a research intern—and taking in the many kinds of things being named and related as systems—enabled me to examine both the technical and the cultural procedures by which systematicity is constitutive of U.S. social and political life.

## Analogical Systems

After our boat pulled up to the life-support buoy on NEEMO 12 splashup day, everyone on board squinted through blazing sunlight to see crew members' familiar heads break the ocean surface after the swim upward. Wet-suited technical divers helped the swimsuited returnees haul themselves onto the boat. The aquanaut crew of four NASA personnel and two support technicians from the research center gave dripping thumbs-ups, coughing, laughing, and waving. I came to know such crews through the interacting socialities of research—mine and NASA's—and in demographic terms common to research disciplines. NEEMO continues to be part of NASA's international spaceflight training partnership, so its missions include personnel from Japanese, Canadian, and

European space agencies. NASA spaceflight crews were, during my fieldwork time, mostly—but not all—white and male, with the majority U.S. citizens, like the average crews in the International Space Station and space shuttle programs. Historians who have documented NASA's 1960s Cold War and civil rights–era efforts to integrate its workforce argue that crew social composition came to serve two institutional purposes: first, to assemble expertise, and second, to represent governmental control of social inclusion in what those historians and their research subjects both refer to as systems—technical, political, social.[10] The splashing-up NEEMO 12 participants were all U.S. citizens organized in a typical spaceflight hierarchy of command and technical status: a white female veteran astronaut mission commander, a Latino male astronaut trainee, a white male physician astronaut hopeful, a white male aerospace flight surgeon, and two white male undersea research center habitat technicians. On the boat, they took their first breaths of nontechnical air. Back at the National Undersea Research Center's mission control room in a Key Largo condo living room, public affairs staff members were sending press releases about the crew's safe return to networks of participating groups, which included researchers, media outlets, and classrooms of schoolchildren.

I have begun with NEEMO 12's splashup, but in the rest of this chapter I recount my fieldwork in NEEMO's longest mission to date: NEEMO 9. I roll out this ninth analog mission's metasystemic processes, starting with situated analogizing, as I learned and encountered them in research and management sites from underwater to outer space.

With the development of space programs in the United States and the Soviet Union came places on Earth claimed to be "space analogs." As analogies in action, space analogs serve national purposes as parts of space programs and centers. On one hand, they are simulacra in Jean Baudrillard's sense: versions of a thing that are no less socially real than what they represent.[11] But space analogs are more situatedly elaborate than technical simulations and are meant to partially reproduce space's environmental conditions: research subjects lying semiprone in hospital beds as if weightless in space, people working in spaceflight-like scuba gear in indoor pools, animals and humans in civil and military spaceflight simulators, research teams in desert and lava-bed campsites, test "crews" in isolation chambers, and trainees and researchers living in undersea habitats. These space analogs are managed as parts of everyday known and imagined *un*known government and industrial work sites. When experts demonstrate the analogousness of environmental and systemic forms and functions in these sites, they often extend the analogies into symbolic and speculative futuristic domains. This imaginative analogizing applies to other kinds of U.S. space-like places, from early twentieth-century world's fair rides to contemporary theme park sites like Disneyland's Space Mountain to natural sites metaphorically likened to outer space and given names like Craters of the Moon National Monument. These constitute a heterotopia of "space-analogous" sites on Earth. Analogizing can make such heterogeneous parts cohere into systems, connecting functionality and purposes and ordering the social and material relations of a current and future environmental space.

As a result, space analogs put space programs and systems into speculative astronautical geographies *and* ecologies. In 2005, three months before the *Exploration Systems Architecture Study* was released, NASA

Aquanaut/astronaut Ricky Arnold winds up the Aquarius habitat's "umbilical" air lines. "Infinity symbol, cool," a student lab technician remarked to me when the photograph was posted on the NASA NEEMO website. Courtesy of NASA.

administrator and systems engineer Mike Griffin used a now-notorious analogy to introduce a new lunar exploration program: "Think of it," he advised news media and publics, "as Apollo on steroids."[12] Along with its attention-getting comparison of a space program with a Greek god shot up with illicit performance-enhancing drugs, the analogy points to systems work as a vitalistic process of perpetual technical improvement. Today, "retired" NASA space programs like Apollo are institutionally understood as the agency's "heritage systems," where "heritage" signals an ongoing redevelopment of systemic things and cultural processes that began at midcentury.

U.S. systems engineering emerged as the Cold War–era alliances of military and industry funneled experts into the theoretically endless horizons of national spatial control projects, from long-range missile and satellite systems to the transportation infrastructures human spaceflight would extend beyond Earth's surface. In this way, the "systems approach" became as socially portable as Frederick Winslow Taylor's scientific industrial method. Historians claim that while system practice has a hard-to-trace history, its mid-twentieth-century boom can be traced to events advocating systems thinking, such as the well-known Macy cybernetics conferences and the less examined meetings of the Society for General Systems.[13]

System practice also circulated as a form of organizational procedure, including the formal and informal analogizing that builds systematicity. Among the many historical project manuals I found in NASA archives was a technically masculinist-inflected 1973 handbook titled *Project Management in NASA: The System and the Men.*[14] In such settings, as engineering historians Thomas P. Hughes and Agatha C. Hughes describe, the "holistic vision" of systems as sets of cohering "heterogeneous components" was "spread by analogy."[15] This is the same systemic vision that built up structuralist and functionalist thought by imagining systems as universalized forms organized by function. As a result, biological and technical things like bodies and machines became imaginable not only as structurally and functionally analogous but as dually manageable under the command of a larger purpose. Although this work happened in various kinds of systems-focused work sites, the aerospace industry's extremely extensible analogical spatiality and temporality gave this alignment of form, function, and purpose an expansive quality of infinite scalarity.

The centralized governance of U.S. aerospace systems began with NASA's precursor, the early twentieth-century National Advisory Committee for Aeronautics (NACA). Until it was dissolved in 1958, NACA brought European and U.S. industrial warfare experts together to build technical and organizational systems aimed into the cosmos. After World War II, NACA integrated German scientists into U.S. military and research organizations through a program called Operation Paperclip, reterritorializing militarized European systems management into homegrown U.S. versions. Later, under the leadership of figures like German rocket scientist Wernher von Braun, who also worked to produce public science and spaceflight propaganda, NASA contributed to the technical, spatial, and social consolidation of U.S. "big science." From 1958 to 1962, NASA established space centers to coordinate its strategic sites along coasts and in deserts and to assemble multidisciplinary programs.

From the Eisenhower era onward, large-scale systems management shifted from military to civil contexts. This transfer was animated by new cross-disciplinary concerns about the control and management of built and natural complexity. So-called complex problems required responses in terms of part/whole structures. Through the 1960s, formal collaborations among engineering, information, and biological systems practitioners became common in institutional and corporate sites across the United States, and universities graduated their first systems engineering classes. During this time, systems-oriented experts moved in and out of academic and government settings, including NASA centers and centers contracted with NASA, such as the Jet Propulsion Laboratory in Southern California, affiliated with Caltech. In 1971, JPL hosted a who's who of U.S. systems approaches in its lecture series Systems Concepts for the Private and Public Sectors, the collected contributions to which were published in 1973. In his lecture, JPL director William Pickering used biological metaphors to describe the "evolution" of JPL's adoption of the systems concept to ensure "an integrated and optimized system," requiring a "matrix organizational structure" to coordinate the "functional boundaries" of subsystem elements.[16] Military contractor Simon Ramo contributed an avidly expectant lecture titled "The Systems Approach," with the subtitle "Wherein a Cure for Chaos Presents Itself."[17] Among the contributors to this volume, chaos was understood to threaten all human-made and -controlled systems, from sewer systems to gambling systems and even to modes of "beating the system." In this systems thinking schema, all kinds of systems are analogous if they meet the definition of "a set of concepts and/or elements used to satisfy a need or requirement" determined by bureaucratized hierarchies of systems experts in the service of national interests.[18]

When I entered the spaceflight institutional field at JSC, modes of anticipatory and functionalist analogizing were active and widespread. JSC engineers and scientists were preparing to build new systems guided by the production of top-down functional requirements tied to the presidentially mandated redevelopment of heritage forms. At this point NEEMO was moving from being an analog for space shuttle and space station missions to one relevant to interplanetary missions. But throughout this time NEEMO also operated in the context of insecure rationales and funding for human spaceflight. Even as human spaceflight faces criticism as a luxury expenditure, its off-world systems justify and are justified by developmental logics of environmental expansionism linking technological heritage and colonial pasts with transspatial technological futures.

The Johnson Space Center's very landscape, when approached by car, analogizes colonial and technical systems building and their roles in the governance of U.S. human–environment relations. To get to the JSC campus from the freeway, you pass a sign marking the entrance to the city of Webster, which declares itself the "gateway to the future"; you drive down to Saturn Boulevard, then up to JSC's front gate. There, employees and visitors are greeted by a frontier heritage analogy: a herd of Texas longhorns grazes next to rockets in JSC's publicly accessible Rocket Park. The scatter of impassive cattle, descendants of those brought to JSC in 1997 in homage to the settler colonial past and to boost local students' participation in center activities, adds a whiff of nation-building history to the Gulf Coast air that blows around the stark black-and-white rocket cylinders. This scene analogizes two American ages, two kinds of energetic power, and two kinds of modern spatial dominance: territorial and environmental. As symbols of territorial expansion and adaptability, rockets and cattle signal the importance of controlling habitable conditions, from western ranches to southern air-conditioned buildings to pressurized space capsules. Historians of the Houston area join global colonial history scholars who describe the connected symbolic and political economic roles of environmental control, from energy production to swamp draining to air-conditioning, in the governance of colonized spaces deemed threatening but also adaptively vitalizing for European bodies and sensibilities.[19] The cattle/rocket park also illustrates historian Frederick Jackson Turner's well-known nineteenth-century hypothesis of the American frontier as territorially finite but nationally vital by showing that frontiers live on by analogy.[20] Further, it indexes President Kennedy's 1962 Rice University speech in which he likened sea and space, describing Houston as a western-turned-scientific "outpost" of a "new frontier" and justifying the lunar program as something worth undertaking because it would be "hard" to do.[21] All in all, the scene shows visitors a Cold War cosmology being made up through territorial conquest and extreme environment engineering.

From the moment I began fieldwork within JSC, I became immersed in daily productions of analogies linking spaceflight functions to national development purposes. My first interlocutor, systems engineer and program manager Curt, occupied an office on the top floor of the central administration building. Another interlocutor analogized this central building to "Mount Olympus" and the nearby Astronaut Office to "Valhalla." Curt started at NASA at age nineteen, in a co-op program that places college interns in its centers. He is a living example of the agency's explicitly system-focused political mandate to compete with the Soviet system at a time when, in the words of 1958 Space Act analyst Glen Wilson, "the status of our educational system" became a "serious concern."[22] The view of buildings and grassy swales from Curt's office was calm and orderly, belying his endless struggle with adverse physical environments and dicey agency futures. About to retire, he would later go on to SAIC (Science Applications International Corporation), which, like the RAND Corporation, is a government defense contractor and distributor of systems thinking. Curt explained that he took this final JSC management job to "control" the "public relations environment" so that human spaceflight would be accepted as vital to the nation's future. He had witnessed three decades of attempts to justify human spaceflight in commissions, panels, polls, and consultant reports, some of which he gestured toward while he talked about what he termed "strategic positioning" problems:

> We didn't use [the end of the space race] as an opportunity then to move beyond the technical to figure out why, why should this country really send people into space? Even on the rare occasions when we did, it was just in rhetoric. Well, "we should because destiny of the species, yada yada yada, yada yada yada." It was fodder for speech. It never took root in building a structure that actually shaped what we did around what we concluded was the reason to do it.

Curt went on to make analogies, as did other interlocutors, between spaceflight and railroad or highway building, highlighting the systems engineering perspective that justifications should be built into the things that evince them. Railroad building is a project that historian of American industry Alan Trachtenberg argues was explicitly described as a symbol of the ascendency of a "city and factory system" over Native American "communistic" systemic practices.[23] Although Curt was reflexively critical of "just rhetoric," including manifest destiny rhetoric, he, like others, often spoke to me of space exploration as a form of evolutionary progress. During my fieldwork, engineers and scientists as well as administrators used the words *human* and *species* in ways that effaced the racialized and gendered histories of these terms and, therefore, of system as a sociopolitical concept.

The systems engineering rationale that technical things should argue for themselves was built into the NEEMO analog program, which follows a half century of analogically entangled NASA reasoning about why human life should be extended into the solar system, what the agency should build and do, and how it should connect reasons with actions. Despite the bad reputation of analogizing as an inferior form of scientific reasoning, space workers, like other experts, depend on analogies in work that involves huge gaps in time and space and between knowns and unknowns—in technoscientific as well as social domains. Analogies made to argue for space exploration not only make simple one-to-one comparisons but also often interrelate all kinds of progressive processes, such as prehistoric migrations, imperial "voyages of discovery," and childhood development. Such scientific analogizing can be used to justify comparisons driven by social convictions that lack empirical veracity, as is evident in the role of analogizing in the history of scientific racism and sexism rhetoric, in which human difference has sometimes been analogized to human evolutionary development.[24]

Supporting such social and evolutionary analogies is Apollo astronaut Neil Armstrong's famous form–function metaphor, made as he set foot on the lunar surface in 1969, equating the "step" for "a man" with a "leap" for "mankind" in technical space and time. Theorists of analogy describe how experts situate and compose acts of analogizing in order to make jumps or "leaps" of reasoning from known to unknown.[25] Spaceflight analogs like NEEMO enable analogical leaps between the systemic forms and functions of organisms, machines, and environments. In this way, spaceflight analogs exhibit what feminist methodologists, citing Donna Haraway's analysis of the impact of cybernetic information theory on biological thought, have termed the "metaphysics" of postwar functionalism.[26] The NEEMO analog, from its position undersea, makes an operational "argument ecology" for the value of spaceflight's leaps within a larger cosmic ecosystem.[27]

## NEEMO 9: Making Analogical Ecosystems

During the planning of all three missions in which I participated, I spent time in the NEEMO "bunker" at JSC, a small room that holds equipment and documentation, serves as an organizational center, and displays its programmatic identity. The bunker is tucked off to the side of a mazelike hallway within one of JSC's classic midcentury concrete buildings, which contain well-worn original fixtures, laminate flooring, and Formica countertops. I went in and out of this bunker and into the other JSC meeting rooms and off-site spaces where NEEMO mission activity planning occurred. All NEEMO missions feature a dizzying array of activities linking NASA with its partners and participating allies: contractor companies, universities, other government agencies, the military, schools, media outlets, NGOs, and individually networked colleagues and friends. NEEMO crews, whose photos appear on the bunker's walls and in filing cabinets, include astronauts, "as-cans" (astronaut candidates who have not yet flown), "as-hos" (nonastronauts who are hopefuls), and technical experts from various NASA divisions. All of these kinds of NEEMO participants engage in what they call the "expeditionary" testing of equipment, bodies, and processes linked to the formal system of systems architecture of the Constellation program. As an intern hired by the National Space Biomedical Research Institute (NSBRI), housed in Baylor College of Medicine in Houston, I entered this partnership network during the planning of NEEMO 9, a training mission that was typical but longer and more technically complex than any previous NEEMO missions. It was tailored to the interests of NASA's partnership with the Canadian Space Agency, and the commander of the mission was a white male Canadian astronaut. It also included two American astronaut candidates, one a white male and the other a white female, as well as a white male physician and astronaut hopeful and the requisite two habitat technicians supplied by the research center, young white males with extensive technical diving experience. During the planning phase I accompanied my supervisors on trips to Pennsylvania, California, and Canada. NEEMO 9 was scheduled to field-test two kinds of interrelated biomedical technologies: U.S. human performance technologies and Canadian telemedicine technologies. Both kinds of technologies are promoted by governments and businesses as systems that improve environmentally situated bodily systems and as techniques that accelerate social adaptation in environmental extremity. The technical systemic analogizing of this mission would create an unusually extensive space of emerging relationality, invoking the scope and scale of spaceflight's ecosystemic space.

The NEEMO bunker connects material and symbolic elements from decades of governmental programs to control human ventures into sea, atmosphere, and space. The Aquarius habitat was built in 1986 and proclaimed by the National Oceanic and Atmospheric Administration to be "the world's only underwater laboratory [or] habitat." Aquarius was part of a strategy to make NOAA the "wet NASA," but it became part of NASA operations as well. Topological maps of Aquarius's surroundings displayed on a bunker wall and stored in drawers show that Aquarius is located in the "Scott Carpenter Basin," named for an Apollo-era astronaut who became involved in marine–space analog projects. Other images in the bunker communicate NEEMO's liminal programmatic place in an outer space agency. The astronaut trainers who invented NEEMO, Craig and Lance, told me how they try to make the program as multidimensionally valuable as possible as a hedge against accusations that NEEMO missions are vacation-like "boondoggles." They intentionally chose a name for the program that would result in an acronym geared for broad appeal, evoking Jules Verne's heroic submariner antihero Captain Nemo and his nature-avenging crew, as well as that character's nonhuman cartoon namesake, the Disney/Pixar hero Nemo, a clown fish who takes a multispecies adventure to return to the sea after being held captive as a pet on land. The bunker also stores a pirate flag, which staff members bring to Florida during missions—not as a signal of the territorial claims marked by "flags and footprints" in spaceflight but as a symbol of NEEMO's claims to outsiderness.

NEEMO's official program graphics and mementos, stored in the bunker, also claim a metaphysical and evolutionary systemic form. Like most U.S. space program workers, I collected a lot of program swag, including a T-shirt with an original NEEMO program logo. In this logo, a ship porthole frames a yin/yang shape illustrating scenarios technically theorized as complementary: an Aquarius-with-aquanaut drawing is set in the light-colored "watery" yin half, and a space-station-with-astronaut drawing appears in the dark "fiery" yang half. In this analogical enthymeme, the cyborgic aquanaut/habitat system is also metaphysical. It asserts that human ecological adaptation to more-than-earthly life is transcendentally meaningful as well as purposeful.

As NEEMO founding civil servant Craig explained to me during a NEEMO 9 planning meeting in the bunker one day, *all* spaceflight systems should be regarded as analogs for subsequent ones; this makes them already developmentally analogous to as-yet-unbuilt future systems. "That's why we call ISS [the International Space Station] an analog . . . we have [also] used the term 'Mars forward.'" This phrase is used for activities that appear to push space program development toward Mars in time and space. The idea that human destiny lies in mastering survival in a succession of progressively more extreme sites exemplifies what David Nye calls an American "second creation" foundational narrative, in which new sites of occupation and technological development are regarded not as final destinations but as springboards from which to engage the next productive site.[28] In the NEEMO bunker, a poster commemorates early saturation dive projects that aligned technological development, adaptable bodies, and a new perception of what counts as human space. In the mid-1960s, the first French aquanaut, Robert Sténuit, expressed the cosmological perspective of those projects: to open up a "second universe" in which "the improbable is normal" and in which new environmental resources, both material and spiritual, provide a "guarantee of the future."[29]

For months, in the bunker and in other sites, I participated in prepping for NEEMO 9's experiments to test whether biomedical technical systems could be adapted to work in an undersea habitat's extreme conditions. This was also work to validate such adaptive processes as futuristically productive parts of national systems. Funding for such NEEMO activities came directly from NASA, distributed extramurally through institutes like the NSBRI, and from other sources that fund researchers to participate in NASA projects. One of the aims of the NSBRI during this time was to "provide medical monitoring, diagnosis and treatment in the extreme environments of the moon and Mars" and then "transfer" those *Star Trek*–like innovations to Earth for patients "suffering from similar conditions" as astronauts, such as "osteoporosis, muscle wasting, shift-related sleep disorders, balance disorders and cardiovascular system problems."[30] Through the social loops of spaceflight funding and analogy making, the systemic political economy *and* political ecology of spaceflight research converge.

One such connective loop was evident in the planning of NEEMO 9's showcase event: a robotic telemedicine demonstration meant to analogize the interconnected future of astronaut health systems, battlefield health systems, and the Canadian health system's reach into remote and extreme spaces. In this experiment, a surgeon in Canada would remotely operate a telemedicine robot located in Aquarius to tie surgical knots in a piece of foam in the habitat "as if" doing surgery on an aquanaut or astronaut. This seemingly simple act both continues and updates midcentury heritage systematicity. During the mission's planning phase, I went with my NSBRI coworkers to the Center for Minimal Access Surgery in Hamilton, Ontario, where director Dr. Mehran Anvari and colleagues were testing the feasibility of deploying the da Vinci Surgical System robotic arm in extreme conditions like the wet, pressurized, and cramped environs of Aquarius. The U.S.-based company that invented the arm, Intuitive Surgical, gives the device a robotic genealogy that relates Leonardo da Vinci's fascination with knightly armor to science fiction author Robert Heinlein's story about a disabled man and his prosthetic remote manipulator.[31] Both Anvari's center and Intuitive Surgical promote the promise of telemedicine to collapse the physical and social distance between doctors and patients. This set of technical links creates metasystemic relationality among this NEEMO surgical demonstration, civil and military medicine, national and extranational space, and the current and future governance of environmental systematicity. The aquanauts and Anvari were set to narrate these relations to publics during the mission, giving systematicity an evolutionary narrative made charismatic in the extreme setting.

NEEMO 9's telesurgery experiment sutured foam rather than flesh, but it demonstrated NEEMO's charismatic role in analogizing technologies that dramatically extend the coordinated management of environmental bodies and spaces. Throughout training and missions, I participated in supporting NEEMO aquanauts to use their bodies as environmentally responsive instruments to generate medical data on extremes while they also engaged in exploration-analogous work gathering data on their alien environs—a task they embraced as part of the value they personally placed on work in extreme environments. Aquanauts'

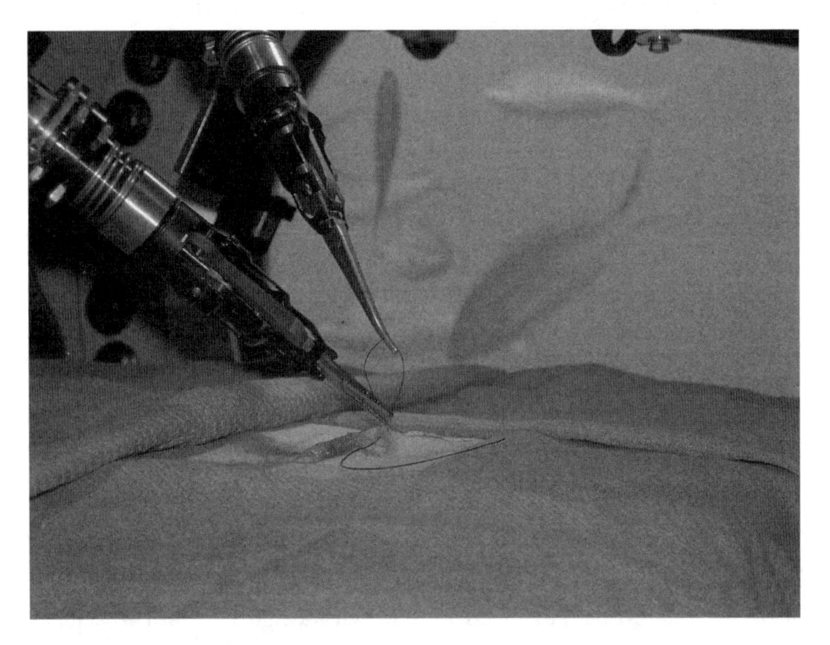

A spaceflight analog "first": telemedical surgical knot tied underwater on NEEMO 9. Courtesy of NASA.

self-monitoring tools are designed to enable control and optimization of "performance" underwater, in outer space, and at earthly extremes. The NEEMO aquanauts were trained to engage in frequent bouts of psychological tests, to use wrist-worn technologies to measure sleep/wake patterns, to use push-button devices to measure eye/hand reaction times, and to don wearable systems, as I describe below, to capture a range of physiological measures used in biofeedback.[32] All of these technologies were to be NEEMO's contribution to new spaceflight behavioral health management "subsystems" that included embedded "unobtrusive" face- and voice-monitoring devices and a "computerized 'psychologist'" for astronaut self-diagnosis and treatment in extreme conditions.[33] But NEEMO aquanauts were doing systemic medical self-monitoring while also engaging in systemic ecological data collection; mission photos often show medically monitored NEEMO bodies engaging in activities mimicking off-world scientific exploration while also generating real ecological data, such as by conducting coral reef biomonitoring. This systemic comonitoring of bodies and environments extends what feminist philosopher Rosalyn Diprose describes as the "biopolitical technologies of prevention and pre-emption" across normal and extreme spaces.[34]

NEEMO is one among many U.S. sites producing technologies of the quantified self, not just as a normalized biological subject but also as an ecologically optimized subject. During NEEMO 9 planning, I went with NSBRI staff to a lab at the NASA Ames Research Center to learn how to train aquanauts to don and doff a wearable cybernetics-inspired body-monitoring technology. The Autogenic-Feedback System (AFS), designed by Dr. Patricia Cowings, was a biofeedback device designed to make the wearer transenvironmentally adaptable. It consisted of a complex harness with sensors precisely affixed to the skin of a finger and the chest to transmit data on vital signs, from heart rate to respiration rate, and a storage card from which data were uploaded onto a computer. Data recorded in the Aquarius habitat would be evaluated to see if they were analogous to data recorded in spaceflight. Cowings explained her work as "teach[ing] people to control various bodily functions," including space sickness symptoms, by responding to bodily data. She came to NASA as a biofeedback scientist and became one of its first female African American astronaut trainees, although she was not selected to fly. Her biofeedback research is historically associated with appropriations of cybernetic theory by 1960s popular psychology, but its principles are now ubiquitous in the form of wearable fitness devices and medical monitors. Such devices make people responsive to the relationship between environmental stimuli and normative health signs, like blood pressure, usually by engaging in bodily control practices, like breath control, or by taking substances. Later during this training trip, in another lab setting in another institution, I spoke with a female ex-astronaut engineer who was working on a small portable biomonitoring device for military applications. She described its usefulness for enhancing "situational awareness" in battlefields and in space: "After all, we're all concerned with the same thing: extreme, remote, and secure."

Technical systems that analogize environmental conditions in space to those on Earth give metasystemic dimensionality and a sense of ecosystemic relatedness to processes like civil and military health care delivery, sleep monitoring, and interactive body–environment adaptations. Another result is that the extreme risks and costs of spaceflight systems seem spatially as well as temporally purposeful. The idea that spaceflight is functional in immediate social and putatively species-level ways connects the scalarity of progress and evolution, a connection that also plays out in systems time.

## NEEMO Mission Timelines within Systemic Project Spirals: Making Processual Temporality

What is systems time? Let's start with the result of mission planning: the launch of NEEMO 9 on mission day one in a dockside research center in Key Largo Florida. On day one, I wear a T-shirt given to me by JoAnne, a young space station timeline scheduler, that describes NEEMO as "exploration from the bottom up." Timekeeping control finds its apogee in spaceflight programs. I had arrived in Key Largo the day before, and I am now standing on the dock outside the National Undersea Research Center's research building complex of converted vacation condominiums. I will spend the entire mission here helping to manage the biomedical data coming in test tubes and in downloads from devices that are "potted up" from the habitat by divers each day. The dense, humid air is a tangy blend of brine and diesel fumes as staff and aquanauts push scuba gear onto the slick deck of the center's small, well-used boat, *Research Diver*. The slate-blue sky is cloudless; sunlight blazes. Those of us who will be landlocked "topside" in the condos exchange damp hugs with the crew we have been supporting for more than a year. After the crew boats out and reaches the Aquarius habitat, we will see them on a daily video feed as they proceed, as mission planners would say, to "execute" the eighteen-day schedule. One former NEEMO aquanaut called it the "almighty timeline"—a label that referred to its ordering power within missions and to its transcendental significance for systems practice.

NEEMO 9, now in action, demonstrates how systems time is geometric and linear only to a point, beyond which it takes on naturalistic forms that perpetuate the vitalistic scalarity of human/machine systems coevolution. Scholars of robotic spaceflight systems work have shown how Mars mission work creates temporal clashes between earthly embodied and otherworldly machine time.[35] Managers have to coordinate the day/night "cycles" of humans and solar-powered vehicles with linear event-based plans, which results in overlapping nonlinear human/Earth/Mars schedules that act like solar systemic time zones. This coordination evinces a metasystemic mode of processual time that is biomorphic: timelines are programmatically nested within models of environmentally driven planetary exploration "life cycles" that fit within a "spiral"-shaped model of technical systems program development. In robotic and human spaceflight, mission timelines, systems development practices, and visions of unfolding human evolutionary futures in large-scale environments are vitally aligned.

Next to me on the dock on day one stands a lean, tanned, white middle-aged veteran astronaut, Jason, who, as a NEEMO science activity principal investigator, has a set of precisely timelined activities to monitor. I have not met him previously, but I know that in a former career he was a technical diver for an oil exploration company, and now he designs space analog tests to improve the "centers of gravity" of environmental control backpacks for use on future spacesuits. Jason's team will put prototype life-support backpacks on NEEMO 9 aquanauts, who will walk around in weighted boots on the rolling knolls of the coral sand bottom near the habitat, enacting the data-producing analogy of what encumbered bodies feel and act like on planetary surfaces. Perspiring from the strenuous activity of organizing equipment, he turns to me and asks: "Who are you? You don't look like NASA." I tell him, "I'm a research assistant, and an anthropologist, here studying the science and technology of spaceflight culture." He responds, "Well, this isn't a science and technology culture, it's an *operations* culture." Furrowing his brow, he gestures to all the dive equipment stacked next to the backpack prototypes-in-development.

Jason's statement reminds me that analogs are scientific second to being "operational," a military systems work term signaling how programs themselves are imagined to function in space and time like machines.[36] This machinic image fits with the technically precise lines and grids on the now slightly damp copy of the mission timeline I hold in my hand. In the working fever of day one, everyone is now, in NASA-speak, "working the timeline" as if making a machine work. But in the context of system-of-systems practice, the program's temporality exceeds the mechanistic. NEEMO is also considered to be part of a larger, more organismic model of "evolving" technological systems. The idea that technological change is not just a by-product or factor of social change or human evolution, but rather technology itself "evolves" in relation to environmental factors, including human selective action, is a discursive hallmark of systems and operations practice. The idea that technology is vitally emergent appears also in the work of philosophers who do not explicitly engage systems thought but are explicitly influenced by systems theory, such as Bertrand Gille, Gilbert Simondon, Deleuze, Guattari, and Bernard Stiegler, who all elaborate theories of technology and the machinic as vitally emergent forms. In a related way, systems and operations experts also theorize timelining and, ultimately, spiral program development as vital evolutionary practices.

On mission day two, I walk, as I do every day, into the "mission control" condominium's living room, where a Navy SEALs program sticker on the condo's sliding door greets me: "The only easy day was yesterday." The slogan sums up the specific and general idea that extreme expeditions generate personal and social transformation through confrontations with naturally occurring, incremental challenges. I am immediately in the thick of what amounts to operational timeline control, where events, things, and messages are coordinated and disseminated to NASA and beyond its borders. The condo's small mission control/living room is always crammed with NASA astronaut trainers, flight controllers, and a couple of dive consultants. Also present are visiting NEEMO activity partners, both American and Canadian, and various "friends of NEEMO" gathered to witness each day's events. The condo's kitchen counter is piled with scribbled-over timelines in a constant state of revision, notepads, telecom equipment, and bottles of neon-colored energy drinks. A crowd of intently focused managers and technicians often gather around the mission's harried and sleep-deprived scheduler, who sits in front of a large laptop screen, constantly revising a digital timeline. She revises what happened and what will happen, with a technique of foresight that is more like forepositioning, as plans are made to move people and things around in order to "stay on time": readying aquanauts for dive excursions outside the habitat, for media interviews, for biomedical experiments.

In between NEEMO missions, I read spaceflight system-of-systems planning documents at JSC and NASA Headquarters, finding not just a universe of spiral galaxies but a world of project spirals. The spiral project model is a graphic of project development that represents "progress" not as a time*line* but as a spiral that emerges from the center of an axis and circles outward, passing repeatedly through certain developmental activities, such as testing, review, and execution. The point of a spiral systems design process, as engineers and life scientists told me, is to "retire risk" or to avoid problems like "death spirals" of accumulating problems due to inadequate reproduction of processes of review and revision. Missions and their timelines are serial events folded into such spiral models. These biomorphic evolving "spirals" of systems project time management represent a developmental shape of Western temporality. Nancy Munn advises anthropologists to examine the enacted cultural forms that "temporalization" takes as the result of practices she refers to as "projects" organized around shapely connections of entities and things in space, a process Stefan Helmreich describes in his analysis of how evolutionary biology diagrams change shape in response to new practices for determining systems of relatedness.[37] Temporality theorist Johannes Fabian, a critic of his own systems theory training and anthropology's allochronic delegitimation of non-Western temporalities, notes that the "system" concept has "always served as a figure of thought related to Time."[38] Although Fabian does not investigate systemic time, he points to how Western temporal representations rely on diagrams to

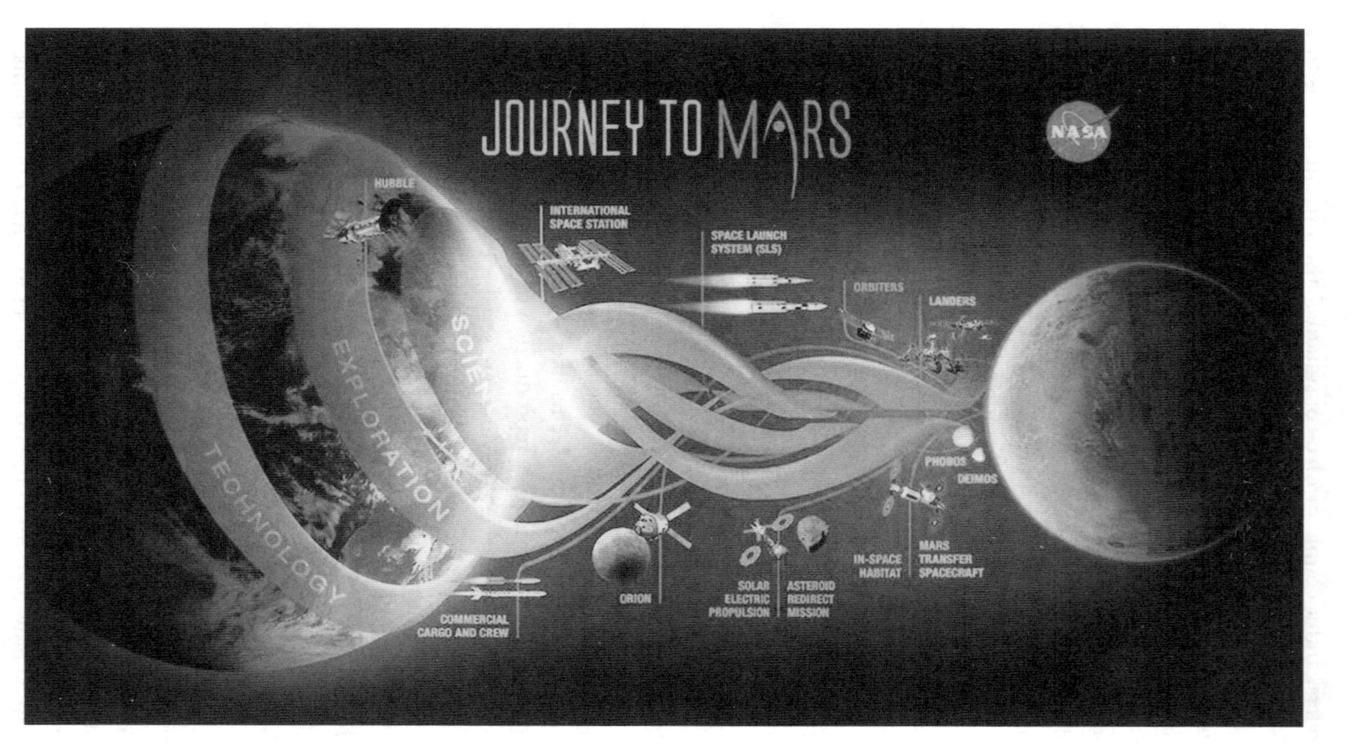


With the stylized *A* analogizing NASA and *Star Trek* missions and linking past and future technologies, this illustration displays the engineering logic of "spiral" design processes. Courtesy of NASA.

naturalize an evolutionary "political cosmology" of space/time drawn in lines, chains, trees, and, in this case, spirals.[39] Critical theorist Elizabeth Grosz has noted how the modern Darwinian "time scale of evolutionary emergence" is a systems-informed notion of life–environment loops of feedback that take the naturalistic shape of a spiral.[40]

In operations settings like NEEMO, linear and spiral temporal frameworks for managing bodies and technologies in extremes are, as my interlocutors called them, "drivers" and "forcing functions" for making systems evolve like living things. Interlocutors spoke of work to map out the "life cycles" of Constellation systems and about NASA's growing emphasis on Department of Defense–approved work applying "spiral" software systems design to hardware design. The spiral process would be a more cost-effective and "flexible" way to fold existing known-risk NASA systems, from solid rocket boosters to software systems, into Constellation. This spiral model takes the ideas of project testing, development, and risk reduction life cycles and prescribes ways to combine old and new ones into a unified spiral model that "evolves" over time. A paper on hardware development lead authored by NASA Marshall Space Flight Center engineer Rebecca Farr in the first year of the Constellation program describes this in organic terms as "growing a system" through modes of experimentation that eliminate "nonviable alternatives." The paper ends with a prediction: "Developing this first vehicle is only the first step in the Journey, it is not our final destination. Those who cannot remember the past are doomed to repeat it, while those who learn from the past, *evolve.*"[41] The "our" here folds spaceflight engineers, national citizens, and generic unmarked humans together into the evolutionary spiral, giving new systems-thinking meaning to historian George Santayana's progressivist dictum.

But in the context of a program design document, it also implies that spacefaring is not just a matter of putting parts together but also a process of evolution marked by hidden social divisions between those who participate in this species-scale project spiral and those who do not. Spaceflight experts spoke to me as if from inside the former group, such as a Life Sciences Division director who told me one day after a NEEMO mission, "NASA's job is evolving humanity," and a chemist at JSC who explained his participation in a plan to develop "in situ resource utilization" technologies on other planets by saying, "I'm a believer in the 10,000-year plan."

## "Being a Part": Participation

Like Curt, the engineer seeking to make spaceflight forms publicly display their purposes, Life Sciences administrator Ken told me one day in a JSC cafeteria that part of NASA's job is to disseminate experiences of extreme environmental difference as forms of public inspiration:

> There has to be a benefit from a novel environment, I mean, think of all the people inspired by scuba diving. So how can removing the one environmental parameter, gravity, that is unchanged for the entire history of the Earth, how can that not be inspiring at the very least, or a change in paradigm? And the people who insist there is no value to anything in weightlessness, in my opinion they have clearly left the path of reason and have entered into the area of emotion or wishfulness. So the whole experiential thing, what is the first thing they ask the astronauts: How did it feel?

NEEMO mission timelines include aspirational outreach: rafts of public affairs and educational activities. During NEEMO 9, groups of schoolchildren tuned in to broadcasts from the habitat and watched mission physicians and scientists explain their work. Video and online media outlets were lined up for aquanaut interviews, from online journals like *Greener Magazine* (with its tagline "The Green Solutions Magazine for American Homes, Gardens, and Families") to network morning talk shows and cable channels specializing in adventure, like Discovery and National Geographic. The mission also included what a CBS News broadcast called the "most extreme phone call ever"—one between the Aquarius and the International Space Station. Opportunities for NASA program allies and collaborators to dive down to see the habitat were coordinated with regular supply and maintenance boat trips. The whole time, public affairs and upper-level managers back at JSC were vetting official text and images being uploaded to the NASA NEEMO website. All of these activities revolved around the process of "being a part."

The classic systemic "part" is as theoretically emblematic as its "whole" but has received less critical attention. The quality of partiality, however, has been critically theorized as an antidote for Western assumptions that value the qualities of wholeness and completeness. Marilyn Strathern and Donna Haraway hold that knowledge about ontological

NEEMO 14 represented as graphic novel cover and participation-wanted poster. Courtesy of NASA.

connectivity is ever partial, contingently arranged, and perspectival, but when empowered by dominant groups to be total and objective, such forms of knowledge produce marked differences that make up social relations.[42] Such feminist and anthropological interventions expose the situated conceits of a modern part/whole cosmology. Anthropologist Gregory Bateson, in a reflexive attempt to retheorize the mind and its ideas as systemically situated within an open "infinite regress of contexts," including the social to the ecological, declares that the "whole is always in a metarelationship with its parts."[43] This is not just a meta-analytic perspective; it expresses a midcentury epistemological shift in the ways that systems-minded scientific and technical experts have come to understand "partness" as a condition of living and nonliving things as well as a kind of subjectivity.

The modern use of *part* as a cosmological *and* social ordering term invites ethnographic attention, especially attention to the kinds of parts that the system concept continues to produce as a relational technology. NEEMO's "parts"—technical and living—are made to conjoin the social and environmental in strategic ways. Those joins are situated within NASA's ongoing state-empowered role as a model agency of "integrated" social and technical systems and, today, of governmental productions of "participatory science" and "part of a whole" systemic subjectivities. As the Constellation program was ramping up in 2006, "participatory culture" researchers and advocates were advising national-level educational authorities to pursue a more "systemic approach" to new media education in order to ensure that all young people can be "full" participants.[44] This raises the question of what forms systemic subjectivity takes. I found spaceflight advocates in and outside NEEMO mobilizing the subject "part" in different ways, sometimes to strategically efface and sometimes to invitingly encompass spectrums of lived social difference, but always to situate a human "part" within authoritatively defined large-scale environmental systems.

As the mission went on day to day, communications from within the Aquarius habitat depicted the NEEMO crew members' embodied vulnerabilities but also their technical roles as parts of interconnected artificial and natural environments. NEEMO communications abound with references to mission "participation" and NASA "partnerships" with other groups, descriptions of people involved to different degrees as "participants," and aquanaut testimonials that focus not on specific roles but on "being a part" of things. NEEMO blogs also contain testimonials, such as "[Being] an Aquanaut was like staying out in the wilderness and actually becoming a part of nature" and "We are also participating in underwater research with some of the best marine researchers in the world. From outer space to inner space." A post by astronaut Danny Olivas during NEEMO 2 in 2002 nests analogously integrated parts:

> As I descended . . . we came upon this mechanical creature at the bottom . . . Aquarius. Living . . . breathing . . . an integral part of the reef! It's really kind of cool when you think about it. . . . Humans studying, in an unforgiving environment, the effect they have on the health of that environment. Rather than the sea rejecting the habitat and the humans who are possibly leading to her destruction, she has welcomed them, making them part of her. Almost as if to say, "Please, come. Live with me and the creatures that make me. Understand me and be my voice to those of your kind." I know it sounds kind of corny but . . . the sea has embraced the habitat . . . she seems to recognize the importance this vessel has in her future.[45]

Olivas defends the analogousness of NEEMO and space missions, writing: "These two are very much siblings whose deeds and results are difficult to appreciate, but whose contribution will help shape the destiny of our species."[46] He renders the analog system into a remediating agent of human rehoming within a maternalized Gaia-like environment. My interlocutors throughout NASA also echoed this urgent salvational dimension of spaceflight, describing space programs to me as technical alternatives to war, "meaningless" consumerism, social complacency, and environmental disconnectedness. Within such spaceflight social settings, people reflexively view their roles as systemic parts in critical and generative ways.

Space program participants have a political history as socially differentiated "parts" of the U.S. space program and a totalizing integrative Cold War "system." One young white NEEMO technician trained in systems engineering expressed, as did others I met at his level and higher, antipathies toward "the system." He spoke about the Department of Defense "blowing through money" in wars while government alleviation of poverty "was not being [done] effectively." "Maybe the whole system needs to be reformed," he concluded. But his concern about being subsumed within large systems was not antithetical to his or others' commitments to making specific systems work better. Historian Fred Turner has written about the utopian postwar projects of countercultural antisystem systems builders who used systems thinking, cybernetics, and established military technical systems platforms to popularize the *Whole Earth Catalog* system-of-systems mentality as well as the civilian Internet. As historian Clifford Siskin points out, "blaming" systems and remaking them are as crucial to modern knowledge production as systems building.[47] However, experiences of blaming as well as joining systems differ according to explicitly racialized, class-defined, and gendered "parts."

Emerging studies of NASA's command-controlled race and gender integrations reveal the agency as a site of contestations over the legitimacy of systems as national participatory forms. Historians Richard Paul and Stephen Moss document NASA's pre–civil rights–era role in altering relationships between the political geography of race and technological development in the United States. In the 1960s, NASA's southern coastal locations made them targets for federal workforce integration programs.[48] NASA's ensuing racial recruitments were represented in statistics and "first black astronaut" mission participation. In actuality, scores of black male and female engineers and scientists worked to build technical and organizational systems that, as putatively nonracialized wholes, actively segregated and, as Margot Lee Shetterly describes in her history of African American female mathematicians, "hid" them.[49]

In NASA, the concepts of "integration" and "system" always operate with double meanings: social and technical. This calls attention to how modern "systems" obtain, as historian of the human sciences Nahum Dimitri Chandler argues, the legitimacy and power to reinscribe their limits in relation to what he attends to as the "problem" subject that "the Negro" and African American present.[50] Oral histories of black professionals atomized within World War II and postwar aerospace centers detail their experiences as agents of social and technical systems integration. Southern-born engineer Morgan Watson, who worked on rocket heat shields at NASA's Marshall Space Flight Center in Alabama, told

Paul and Moss about the regional differences among his black colleagues in how they would "tolerate the system" of unequal pay, treatment, and advancement between black workers and their white counterparts.[51] Such experiences occurred in a context in which political leaders explicitly described the stakes of the Cold War as the future of "systems": modern, governmental, capitalist, and racially ordered.[52] A question for the social history of system per se, which Chandler's theorization of systemic power invites, is what kinds of partial subjectivities the system concept marks and fosters.

Twenty-first-century government programs, like spaceflight programs, that publicly promote systems thought and participatory science practices formalize a metasystemic subjectivity—a sense of being a part of systems of systems and of their immense scalarity. Similar to what scholars have identified as the "environmentality" of subjects internalizing processes of explicitly environmental governance, NASA's education and public outreach programs can be understood as formalizing environmental *systems* governance.[53] During and after the period of my fieldwork, NASA was actively responding to government mandates to standardize science education and participatory science through new information system tools like crowdsourcing and data sharing. NASA established the Participatory Exploration Office to comply with the Obama administration's "open government" transparency programs, which aimed to increase "citizen" involvement in government science, creating implicit gaps in participatory inclusion between U.S. citizens and noncitizens. The system concept itself is a part of these targeted education and involvement projects. In 2013, a National Science Education Standards report designated "system" as a "core" educational concept.[54] In that document, the Earth system and the solar system are linked as scalar objects, and being scientifically literate is associated with having a sense of systemic subjectivity.

During NEEMO 9, schoolchildren in the United States and Canada participated in events that located Aquarius—and the students themselves—within an environmental system of systems. One event coordinated a video feed from aquanauts in the habitat with a talk by Dr. Anvari, who discussed telemedicine's capacity to collapse technical, environmental, and social barriers, connecting this with the futuristic presence of aquanauts in the extreme and with the future of the children in the audience as potential recipients of such technologies. After describing how some Canadian communities are medically marginalized, he told the viewers:

> Could you have operations at home? Potentially, yes. But our goal right now is to develop systems that allow every Canadian, irrespective of where they live, to have access to the best medical and surgical care. And that same technology will help astronauts explore space, explore back to the moon, and hopefully in the near future, to Mars.

This event presented an argument about technical integration in which the ontological positions of advocates and audience were systemically linked. While giving viewers a virtual tour of the habitat, NEEMO 9 aquanauts compared spacecraft and Aquarius, remarked that fish had become like neighbors, and discussed their childhood aspirations. This kind of "invitational" activity shares with publics a situated experience of systemic embeddedness and portrays normal, everyday environments as spaces between extremes.[55]

During the NEEMO missions I worked on, invitations to participate directly and indirectly were understood to transform public relationships to outer space and the space program. NEEMO produced what managers called "expedition-ready astronauts" but also individuals who have what one astronaut described to me, in aeronautical terms, as technical and social "situational awareness." As one NEEMO manager put it when describing the purpose of public outreach:

> There is a [networking] strategy, absolutely, but it's also something that we nurture right up front. . . . We tell them you're gonna get to do this cool thing, we're gonna ask one thing of you, we hope you become advocates of what we're doing so other people can do it.

That expectation of a conversion-like experience draws together the shared technical and religious meaning of "mission" as "to send out," indicating that participation in a NEEMO mission involves making it connect with other transcendental projects.

My own experience diving to the Aquarius habitat moved me from a peripheral part to a participatory center where the phenomenological dimensions of my own systemic subjectivity come together. One day during the middle of the mission, I am invited to report to the dock right away. Worried about getting seasick, I stick a 1.5-milligram scopolamine patch below my ear and gulp down 25 milligrams of meclizine hydrochloride between sips of Gatorade. With my body and mind already undergoing performance preadaptations, I feel that I am making an embodied adjustment to the sporty, risky, optimized U.S. extreme.

As I haul my gear and tanks of air onto *Research Diver*'s deck to make the trip out to the Aquarius, I sense a shift in my position, aware of being gifted the privilege of systemic self-awareness and self-production. On the boat, I am among elites who consider themselves to be exemplars of new, previously unimaginable ways to survive. National Undersea Research Center manager and dive master Don, skipper Peter, three NASA employees, an NSBRI technician, and a physician diver guest sit or stand amid an impressive pile of gear on the little boat's narrow deck. They represent dozens of tropical and cold-water dives, cave dives, mountain climbs, survival training experiences in forest wildernesses, and as-ho application tests and trials for astronaut selection. As they silently check and adjust their gear, the scene becomes charged with a performative aura. The physician diver on board is a minor celebrity: a microsurgeon and member of the Explorer's Club named Ken Kamler, author of *Surviving the Extremes: A Doctor's Journey to the Limits of Human Endurance,* written a year after he had visited NEEMO to stitch up the injured hand of an aquanaut.[56]

On the boat, I stand next to Kamler, who participated in the production of the NEEMO system's vignettes about spaceflight as a part of human evolution. He documented his dive experience for *Popular Mechanics* magazine, including a story from our dive day in which participants witnessed a huge sea snail being consumed by a sea turtle. He likened the Aquarius habitat to a shell: each is "a product of millions of years of evolution," of ongoing adaptations to environmental risk.[57] Years after this NEEMO 9 article, a Discovery Channel online article about NEEMO 15, as an analog for a human mission to an asteroid, likened the NEEMO crew to a more professional version of the "ragtag crew" in the movie *Armageddon,* whose petroleum-industry mining know-how helps them save life on Earth.[58]

Analogical vignettes linking spaceflight and Earth system futures also connect spaceflight and environmentalist ethical rhetorics about humans as kinds of systemic parts. The "spaceship Earth" metaphor is an example. This concept was animated by astronaut "whole Earth" photographs, which serve as icons for environmentalism, technological mastery, alienated escapism, and an emerging planetary politics in which Western technical authority is metonymically associated with the idea of a governable "global environment."[59] U.S. environmentalist writing of the space age encourages readers to relate to a scalar ecology of parts in wholes: conservationist Aldo Leopold's gravity-propelled "molecule X" traversing the Sand Creek ecosystem, biologist Rachel Carson's birdless morning on a toxically devastated Earth, astronomer Carl Sagan's description of Earth as a fragile "pale blue dot" when seen from Saturn, and ecologist Eugene Odum's description of Apollo 13's space capsule life-support crisis as a small-scale version of today's "spaceship Earth" crisis.[60] This analogy appears in Odum's *Ecological Vignettes,* written to illustrate the "basic ecological principles"—both scientific and ethical—of environmental systemic interconnectivity.[61]

Spaceflight activists mobilize the idea of systemic interconnectivity to justify ecological expansion and belonging. A popular spaceflight idea, originated by the Russian astronautics theorist Konstantin Tsiolkovsky, is that Earth is a planetary "cradle" for humans to outgrow. Another ecoethical justification for experimental systems building, championed by the crew of Biosphere 2, the 1990s American self-contained ecosphere experiment, is that all people are vitally integrated Earth "ship" crew members on the path to knowing and transcending planetary boundaries.[62] The utopian politics of ecological development became a theme in late twentieth-century American science fiction, exemplified in the collection *Future Primitive: The New Ecotopias,* edited by well-known speculative political ecological writer Kim Stanley Robinson.[63] As anthropologist Timothy Choy points out, ecological comparisons create a political geography of particular and universalized ecological problems and risks.[64] Making systems ethically analogous also makes "system" work morally as a relational technology, encouraging subjects to imagine the ethics of belonging within environmental systems.

The transcendental aesthetics of elite spaceflight systems experience are also part of authoritative narratives about social parts and wholes:

from Apollo astronaut and mystic Edgar Mitchell's "noetic" experience of cosmic unification to the observational trope of "borderlessness" noted by space travelers of all backgrounds to the oft-cited characterization of these experiences in space advocacy literature as "the overview effect."[65] Peter Suedfeld, an NSBRI psychologist network partner from the University of British Columbia Vancouver, researches U.S. and Russian spacefarers' experience of "Transcendence (a combination of Spirituality and Universality)."[66] Within his projects, space and undersea habitats act as cosmology-awareness devices. I heard variations of this throughout my time at NASA. JSC chief lunar scientist Wendell Mendell's description of spaceflight as an "epiphanic community" was among the most poetic, emphasizing space science and the calling to it as transcendental. Such spaceflight advocacy portrays maintaining social identities as less essential than being systemically environmental.

In this larger social participatory context, space exploration becomes an extreme form of "aperspectival objectivity" that locates its universalism not in acts of purging of emotion and social specificity from observation but rather in collectively felt assessments of a cosmic human place and purposes.[67] This perspective can naturalize technological inequalities, creating a human evolutionary and transcendental-experiential ladder with spacefaring societies on the top rung. But it also inspires spaceflight advocates involved in social and environmental activism, visible in astronaut Mae Jemison's work to build nationwide STEM (science, technology, engineering, and mathematics) education social inclusion programs. Jemison, the first female African American astronaut, also leads the DARPA- and NASA-funded 100 Year Starship consortium to create a business plan for interstellar human space exploration. As a spokesperson for social futures in space, she appears in the 2015 environmentalist documentary *Planetary*, in which, coincidentally, a NEEMO 9 astronaut also appears. In the film, Jemison recounts her epiphany of being a belonging part:

> The really wonderful thing that happened to me when I was in space was this feeling of belonging to the entire universe. . . . I actually didn't think, here's this Earth, it's the only place I belong. . . . I imagined myself in a star system 10,000 light years away, and I felt, I also belong there. . . . We're as much a part of this universe as any speck of stardust, any asteroid. We're part of this universe.[68]

## *Diving: Participating in the System*

When we get to the Aquarius's life-support buoy, the research boat captain lets us know: it's time to gear up and dive. As I take a stride off the end of the boat and splash down to submerse, I feel I am acting in what historian of Victorian evolutionary narratives Gillian Beer calls a "Darwinian plot" about technosocial adaptation.[69] My lungs extend into hoses and tanks, I notice every phase of my breathing, and my vision opens into a three-dimensional awareness that NEEMO staff would describe as an "underview effect." I kick downward to the habitat with my dive partners, an NSBRI technician and a NASA flight controller trainee, our ears popping as we adjust to a new atmospheric norm. I was instructed during advanced dive training to understand my life underwater molecularly: I exist in relation to portable quantities of air. This is how, aquanauts claim, one can become aware of one's body as an evolutionary artifact of particularities like air, gravity, and atmospheric pressure, but also of the exhilaration of experimenting with these particularities. In a history of how compressed air and beating "the bends" opened up vertical frontiers, John Phillips writes that "the ultimate limits of technological progress will be determined by the physiological limits of the humans who use them."[70] In an extreme environment the body operates under threat, but spaceflight training and public promotion make this sensible, as progress, evolution, transcendence.

As the habitat looms before us like a soda can on legs, encrusted with bright coral life and hugged by shimmering schools of fish, we come upon a timelined activity that calls on the mind to analogize and participate in enactments of envisioning. Aquanaut-astronauts Dave Williams and Ron Garan walk on the seafloor as if on the moon, wearing weighted boots and helmets as if in spacesuits, building a spindly, nonfunctional "lunar communications tower" out of thin white PVC, communicating with each other by comlink, unaware of us hovering to the side. Ron's work with PVC to evince the terrestrial uses of astronautical technology continued after his time at NEEMO in his NGO work to bring parts of space station technologies for water purification to communities in Africa and Mexico. This work came in part from his experience as an engineer and from an epiphany he had while running, he told me, when he noticed the stunning abundance of unused solar energy being absorbed wastefully into the asphalt under his feet. We swim by this analogous lunar scene and pop our heads up into the habitat's wet porch air/water boundary. The smell is impressive: pungent aromas come from accretions of bacteria in dive-gear rinse buckets, collected for experiments on environmentally friendly bacteria-control methods designed by Houston schoolchildren who are enrolled as part of a nationwide effort to produce future scientists. We chat with crew members, shouting over the loud grating and huffing sounds of the air pumping in and out of the habitat from above, reminding me that we are all on life support, but the effect is not uncanny or sublime; rather, it is strangely homey, like being in someone's garage.

Swimming back up to the boat, I put my anthropological self-consciousness on hold in favor of another kind of consciousness: How much air do I have? I experience a temporary muting of ethnographic analytic loops about nation building, elite expenditures and privileges, avenues of technological chauvinism, universalizing slippages and erasures. I remember instead a sensuous passage from aquanaut/astronaut Mike Gernhardt's NEEMO 1 mission journal:

> . . . Floating in the terminator
> In the instant
> Before sunset
> In the moment
> Before sunrise
> In between
> Night and day
> Light and dark
> Life and Death[71]

This passage, written by an astronaut engineer involved in designing new life-support systems, conveys a sense of environmental conditional boundary crossing, of a middle-way relational situatedness between planetary cycles and life cycles.

Days later, after the mission is over, I experience the temporal end of a NEEMO mission, which includes rituals of closure that signify the continuance of participation. Postmission dinner celebrations feature a gift for each participant in Aquarius missions and dives on the habitat: an "honorary aquanaut" certificate, signed by National Undersea

Research Center staff, with an institutionally prescribed environmentalist charge to act as a missionary witness:

> Your journey to *Aquarius* at Conch Reef, in the Florida Keys National Marine Sanctuary, is part of our effort to excite and educate our friends about the wonders of our world's oceans and coral reefs. We challenge you to use this experience to teach others about what you saw and learned.

Working on NEEMO mission cycles, I collected three of these, and I have told, here and elsewhere, what I experienced, as a part.

My participation in the NEEMO technical system was a metasystemic experience. If analysts of postsystems forms like assemblages, networks, and infrastructures refer to systems as nested parts of those relational formations, likewise systems contain cultural modes that connect them to other social processes—namely, analogizing, processual timekeeping and part-of-a-whole subject formation. The next chapter builds on the problem of how systemic participation is institutionally formalized as a matter of enacting the human *as* system.

# 2

# CONNECTION
## THE ECOBIOPOLITICS OF SPACE BIOMEDICINE

"What happens here," Jill, an astronaut candidate, told me in her office one day, "is that you have an Astronaut Office that completely parallels the space program's organizational structure. So that you can have a bunch of bodies that represent all that stuff." To get to the office she shares with two other astronauts, I walked through the JSC Astronaut Office's front entry hall, lined with butcher paper where passersby can write welcome-home notes for returning space station crew members. Then I took an elevator to another floor, where I was buzzed in to the office hallway through electronically locked doors. Jill, a propulsion systems engineer in her early forties, white, California born, works on shuttle payloads and life-support design. And waits to fly. Photos of her two daughters stand on a big metal desk that looks as old as NASA itself. Gulf Coast noonday sunlight cuts through slatted blinds as she tells me about how she wants to be "a part" of the "flow of technologies" from space "down to Earth." She lists some beneficial potential "spin-offs," saying, "Our [earthly utility] bills would be half as much if we just tied a sewage system into the electric company. . . . Bioreactors, you put human poo into the thing, you get methane gas, you can power something, there's nothing complex about it." She goes back to describing how training to do such closed-loop science experiments in space is making her "a better operator." Being a better operator is an embodied process, tied into, as she explains, the systems-based structure of space programs.

For Jill, this involves "understanding your system better," since "any science platform up there is tied into the system side of things."

As one among "a bunch of bodies" paid to represent *and* work on systems as well as to fly into space, Jill represents a kind of modern systemic being. Her description of astronauts' bodies as selected connectors of "all that stuff" is relevant beyond these office walls. Over in JSC's Life Sciences Division offices, streams of data on Jill as a living and potentially dead astronaut refract like light through the sorting technologies of research, revealing the spectrum of information that constitutes her as a spacefaring human. As a research intern at JSC, I worked with both Jill and her bodily systemic data. Such embodied and abstracted social relations, in NASA and elsewhere, are being ordered into kinds of systems—social, technical, environmental.

Ken, a physiologist and administrator who works with Jill and other members of the astronaut corps, deals with them as aggregated sets of biomedical data about environmental limits, tolerances, and signs. In his mid-fifties, Ken belongs to a generation of white males interpellated by midcentury racialized and gendered images of the heroic astronaut. Also an amateur space historian, he is fascinated by the cultural physiology of that figure. He once told me that the halo-like metal spacesuit neck rings visible in official NASA portraits of suited but helmetless astronauts "function" as sanctification symbols. I followed his work group's efforts to turn the human into another kind of transcendent but technical representation: a "human system." Such projects come to characterize the members of particular scientific generations, whose careers, as anthropologist Sharon Traweek shows, proceed in institutional timelines and professional commitments to longer-scale processes.[1] Long ago, Ken's own too-tall-for-a-spacecraft body had excluded him from the vehicle systems specs of astronaut selection. One day, while sitting in a local Mexican restaurant owned and run by an astronaut's family, he updated me on the essential conundrums of human-as-system work.

> Everybody thinks it's a good idea. But I have yet to see it actually happen. I mean, how do you make the human into a system? What is the human as a system? The human can back up other systems, like navigation, but what backs up the human? What is the system backup, the redundancy, for the human, its brain, all that? Part of [the problem] is that we don't know how to specify

> the operating parameters of the optimal human system because we don't have enough diverse data to know, we don't and won't collect data on that, who's that gonna be? How would we know if we were right?

As I detail from here on, "system" may be a taken-for-granted modern concept, but the *human system* idea still is not "black boxed," in the sense that Bruno Latour uses in reference to facts or objects no longer scientifically contested.[2] Work continues on the idea as a discursively vital but institutionally ambiguous problem, as it does elsewhere in contemporary science and engineering.

In this chapter, I examine how human spaceflight practitioners work with the astronaut as a fundamentally ecosystemic, rather than biological, being. I followed projects focused on the human system across JSC's wide, humid lawns and gridded roadways into honeycombed office buildings, further afield into airy aerospace hangars full of spacecraft and spacesuit testing projects, and out into other NASA centers and meeting sites, where the human system is becoming a key connective formation in U.S. environmental practices and governance.

## Systemic Life

Ken told me how he worked for decades studying the effects of weightlessness on mammalian vestibular systems to devise "countermeasures" that would prevent astronauts from fainting or losing their balance upon return to Earth. Like most of my interlocutors, he had strong opinions about the benefits and limits of rendering living beings as risky parts of systems. He described most of his work on this as "making viewgraph slides." Graphs, studies, and PowerPoint slides are such a significant part of what NASA makes that physicist Freeman Dyson once called them "paper NASA."[3] In NASA studies like those Ken worked on, the human system is elaborately aspirational. A fully knowable human system based on comprehensive data on human kinds is a horizonal goal—what engineers would call a "desirement" rather than a fully defined "requirement." Like other spaceflight conceptual formations, the human system is a partly realized experimental tangle of projects, plans, and pushbacks, and it has champions and detractors. Beyond

NASA, biological and ecological discourses in the United States advance similar conceptualizations of the human as a living system. Like astronautics, this work empowers techniques to manage social–environment interactions by characterizing human parts, bodies, processes, and exchanges with other things as systemic. If, as historian Matthew Hersch argues, the American astronaut was "invented" when a super–subject role turned into a long-term technocratic profession, that invented subject was also unprecedentedly systemic.[4] The astronaut is perhaps the most systemic of institutionally defined humans.

What, then, is systemic life? In NASA, the lives and deaths of a small, select group materialize a constantly reconfiguring systemic subject with material and transcendental environmental scalarity. During my fieldwork, the NASA Man-Systems Integration Standards engineering requirements had begun to be revised. The standards constitute a "bible" for users of spaceflight hardware—referred to as a bible at least in part because, as an engineer interlocutor explained, they render

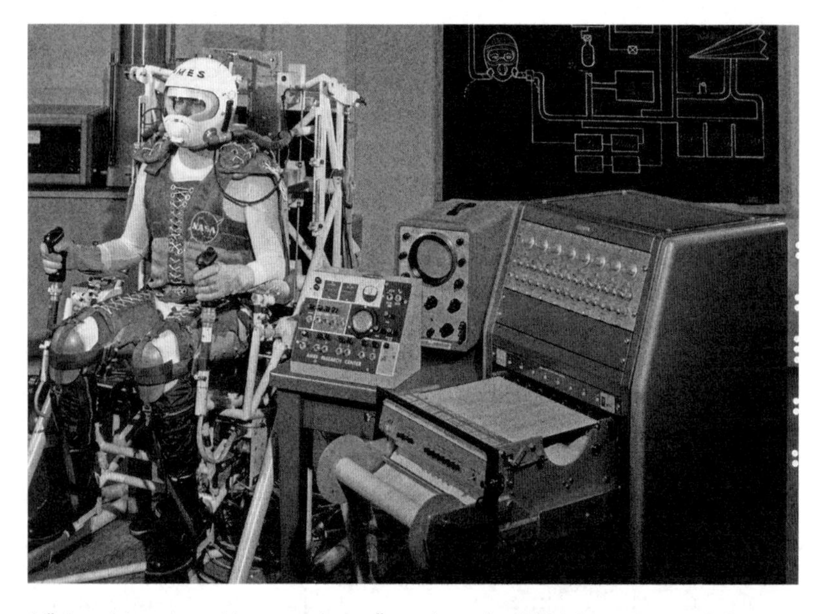


A "closed-loop breathing experiment" conducted at NASA Ames Research Center: the blackboard schematic situates the experimental spaceflight subject within NASA's emerging institutional/technical/environmental system of systems. Courtesy of NASA.

"hard" engineering requirements into "shall statements," like "the Ten Commandments." The spaceflight user had for decades been a "man" fit *within* systems, a man defined by environmental quality specifications, safety and health guidelines, and tables of statistics such as racialized and gendered data about body measurement ranges in the fifth to the ninety-fifth percentile that vehicles must now accommodate due to social "expansion" of the space program's "user population."[5] The latest, 2014 revision of the standards, in which Man-Systems changes to "Human-System," maintains this project but also goes further: by defining the human and crew *as* systems. This "human system" is "to be viewed as an integral part" of vehicle design, "as one system along with the many other systems that work in concert."[6] Both old and new documents claim that human system parameters are yet "unknown" and "insufficient," but what carries across from one version to the next is the relational concept of the "human" as a progressively integrable system, namely, a system of "physiologic parameters" like but not exactly like those of a "mechanical device."[7] People I met who were involved in the revision of the standards explained their use of that mechanical analogy as strategic: they wanted to make the human equivalent to other systems. But in everyday practice, real people are treated as only quasi-systemic in technical and social terms. This work shifts experts' attention from bodies in and of themselves to the broader connective spaces of environmentally situated things and social processes.

I watched Ken and others fit what they called the "squishy" human bodily system within the "hard" engineering systems of the space environment, creating an ecosystemic assemblage amenable to new forms of management and intervention. This terminology of *squishy* and *hard* highlights not only the systems practice goals of connecting different parts and making wholes but also how the partiality and messiness of such connections are phenomena in astronautics work as well as in social theory, albeit in different analytic and political registers. There is a lot of boundary trouble in making actual people systemic. But it is significant that a NASA physiologist was one of my guides. Early twentieth-century U.S. physiologists facilitated the movement of the system concept from biology into biomedical and organizational theory, where "the patient" became a complex systemic subject—both biological and social.[8] During this time, NASA life sciences practitioners began to formalize the human as a duplex systemic subject: a bounded organismal system and an environmentally interactive system. The astronaut in extreme environments has accelerated twentieth-century system work, pushing the envelope beyond the problem of how modern subjects are made socially systemic to how they are made environmentally systemic. System, as a relational technology, becomes distributed within and through the human body and cosmos. This authoritative project exemplifies and extends what cultural theorist Sylvia Wynter describes as the Western liberal "genre" of "being human," which is nested within racialized and gendered scientific "knowledge systems" that construct symbolically differentiated human kinds.[9] Her critical use of "system" illuminates and implicates all systems, concrete and abstract, as political products—including an emerging generic systemic human.

Because the astronautical "human system" is emerging within a Western governmental institution, Michel Foucault's concept of biopolitics, with its focus on scientific processes to control bodily parts and lifetimes, applies here despite its theoretical limits for analyzing the politics of human life in all social contexts. Paul Rabinow and Nikolas Rose's critical clarification of the term reminds readers about its potential for amendment, which the extreme environmental scaling of the astronautical human system concept requires. Rabinow and Rose sum up "the politics of life itself" as the making of "truth claims" based on "knowledge of vital life processes, power relations that take humans as living beings as their object, and the modes of subjectification through which subjects work on themselves qua living beings."[10] They acknowledge that today it would be "misleading simply to project Foucault's analysis forward as a guide to our present and its possibilities"; they call instead for it to guide thinking about the next "mutations" that biopolitical techniques themselves engender.[11] This is a call being productively heeded by social scientists who have identified biopolitical schemas playing out at the molecular level, through the "microbiopolitical" enrollment of microorganisms into regimes of social relations and governance, and the "symbiopolitics" that arise when science presents life as a network of more or less associated beings.[12] But as NASA administrator Mike Griffin boasted in 2007, "What we do at NASA is, quite simply, larger than life."[13]

In a setting where life's spaces are called environments, I show in this chapter how such processes result in "ecobiopolitics."[14] By offering this concept, I call attention to a shift in contemporary loci of control and governance: from life itself to life–environment relations. I describe how this manifests in the systemic lives and deaths of astronauts, which evince both biopolitical *and* ecobiopolitical strategies for managing human–technology–environment interactions. In such strategies, systems are made ecologically interrelational in ways that make it possible for doctors as well as engineers to comanage technical interventions. Governmental human spaceflight, then, becomes an extreme version of emerging practices making ecosystemic governance buildable and scalable from body to planet. This is the ecological space that large-scale systems work affords to power and knowledge. In this work, astronautics practitioners also invest the human system with their own particular hopes and visions for extensively relational futures.

## Astronaut as [Quasi-]System

During my fieldwork I was enveloped in spaceflight's environments, where questions about how to fly things in space exist alongside questions about what "the human" can be. This is not the same as the question of "what it's like to be" an astronaut as a kind of significant public figure, which is a biographical and sociological topic that this chapter sets aside. My fieldwork unfolded through interactions with astronauts as technical biological subjects in formation: as-cans and as-hos, readying to be parts of missions, being systems spokespersons and experts. During NEEMO (see chapter 1), I was also trained to work with "these astronauts" as distributed entities that morph in and out of symbolic and material forms. In the city adjacent to JSC, Clear Lake, streets are named for deceased astronauts, and a McDonald's restaurant is topped by a big inflatable astronaut. Within the JSC campus, astronauts lost in the line of duty are represented by a memorial grove, which is a stop on the public tram tour offered by the nearby Space Center Houston visitor center. The visitor center's gift shop is full of purchasable astronaut things, including children's books with themes like "how to draw an astronaut." Doing my internship, I worked and socialized with astronauts, handled their bodily fluids as research materials, and had to be mindful to keep my germy body away from preflight crews. As I worked with these elite research subjects, who are occupationally classified as radiation workers, I got to know them in complex relation to other experimental bodies, both human and nonhuman and voluntary and involuntary, that are environmentally exposed. The elite, exposed figure of the astronaut spans this and the next chapter as a subject situated within systemic technologies of socially strategic connection and separation.

Spending time with life scientists and physicians in the JSC Space Medicine Division and Human Research Program and with engineers and industrial designers in the Habitability and Human Factors Division, I learned the astronaut's roles as crew member, inhabitant, mission factor, and system in the making. The astronaut fulfills these roles in a work context defined, as Jill's words show, as systemic from its origins. Henry S. F. Cooper Jr. writes about the training of astronauts as managers of individual spaceflight systems.[15] But space life scientists invert this frame by aiming to manage the astronaut *as* a system. Within NASA's complex interdisciplinary organizational groups described as "reporting matrixes," I heard about the promises and the pitfalls of the "human system" idea. Two architects making space habitat mock-ups complained to me, in their sawdust- and foamcore-littered studio, that the human is spaceflight's *most* important system, but thinking about it this way is "totally missing" from design processes. According to a flight surgeon who gives talks on the subject, a fully described generic "human system" seems theoretically desirable but would be "completely bogus" in actual medical terms since it would always be too artificial to be real. A young engineer, a white male former as-ho I worked with on NEEMO, claimed he was glad he had not been selected for the astronaut corps, telling me decidedly, "I didn't want to be a system guy." Such statements point to the technical drive but also the workplace ambivalence about making humans totally systemic.

Like the multifaceted historical emergence of the system concept itself, the NASA "human system" has a complex genealogy as a formal and commonsense artifact of systems projects and policy planning. The authors of the 1959 report of the U.S. Senate Committee on Aeronautical and Space Sciences to President Lyndon Johnson on the Mercury "Man-in-Space" program name the "human system" as a uniquely valuable *non*machine system that advances the socially "practical" rather than the idealistic scientific purposes of spaceflight.[16] The report describes the biologically encoded exploratory "urge" of "man" to "to go where no man has gone before, and to go everywhere."[17] This phrasing links U.S. spaceflight policy with the television series (and eventual franchise) *Star Trek,* which debuted a few years later, and its bio-logic of an exploratory urge continues to be used to justify spaceflight as an extension of human development.[18] It also establishes a strategic precedent for referring to the squishy human as fully systemic but not reductively mechanical. Space biomedicine practitioners who collaborate around this ambiguous human system figure work to make it be, among other things, normalizable across environmentally defined spaces.

## Space Normal Bodies

I was introduced to a variety of efforts to formalize the "human system," but some were more widely supported than others. One such effort was JSC's "digital astronaut" project, which aimed to use NASA's analog and space mission participant data to model the human as a "mission system." One flight surgeon called the fraught and stalled project "idiotic" because its model was not a "natural" human system but one being treated with exercise and medications. Ken and his colleagues were working on another angle. During one of our first conversations, in a big meeting room hung with big, glossy spaceflight pictures, Ken told me, "We need to define 'space normal' so we don't keep trying to treat astronauts in space as if they're sick." Posing normality in environmental terms in order to redefine sickness and health sounded odd, since "normal" is a comparative biological, not environmental, term. But Ken's project, which I describe later, consolidated what I was hearing and seeing: life and environmental processes were being defined as coconstituting. Instead of backgrounding environment as a given or secondary medical variable in the production of normality, space medicine brings it forward and shows how body–machine systemic formations are used to normalize extreme and extraterritorial environments and make them governable.

Michel Foucault's concept of biopolitics provides a tool for understanding how modern forms of normality and health emerge from institutional processes to manage "life itself" at the level of individuals and populations.[19] But this concept only goes so far in the analysis of space biomedicine, which is a branch of environmental medicine focused on managing occupational exposure to extreme and artificial environments. Astronauts are optimally biopolitical subjects, but in a way not fully captured by the concept's anatomy–population axis.

Space biomedicine engages with interactions in spaces where life's environments, whether spacecraft or planets, cannot be separated from life itself. In these spaces, the tension in the duplex organismal/environmental astronaut system model becomes apparent. The politics that result are constituted where space policy mandates meet the institutional politics of treating astronauts as simultaneously impaired by space and super-fit national pioneers incorporating new environments. These politics result from the strategic responses to what Foucault would describe as the "problematizations" of life and health that arise when biomedicine extends into environments that require experimental technologies to support basic survival.[20] Space biomedicine dually problematizes survival and environment, engendering a distinctively ecological modality of biopolitics.

In the next sections I describe the sociopolitical and scientific parameters of the astronaut as a biomedical subject. My analysis is motivated by historian of biology Georges Canguilhem's theorization of "milieu" as a "basic category of contemporary thought."[21] Although aerospace engineering and space biomedicine practitioners use the terms *environment* and *system* rather than *milieu,* Canguilhem's historical analysis of the term *milieu* emphasizes how it conceptually bridges the physical and life sciences in ways that have ongoing philosophical, social, and political implications. Canguilhem traces how the nineteenth-century scientific idea of milieu—the set of relations between an organism and its environment—originated in physics theories about objects connected in space by forces. The concept of a mediating milieu became critical to modern theories of how and why living things interact and change. However, there are periodic "inversions of the relationship between organism and milieu" that influence biological debates about the causes of human variation and evolution and shape understandings of environmental causes of health and disease.[22] The biopolitics of astronautics are environment driven, bearing down on human–environment interstices, not necessarily on human generative processes. This grounds my depiction of what is at stake when space biomedicine buttresses or breaches body/environment boundaries. I describe *ecobiopolitics* and how it works through two signature space biomedical strategies: the invention of the category of "space normal," which simultaneously normalizes bodies and the outer spatial milieus that they inhabit; and the development of concepts and tools for integrating astronauts with systems. I call attention to how astronauts are configured as ecosystemic parts and to the perspectives and experiences of people who work to make that configuration.

By examining how space biomedicine produces cosmically ecological beings, I join scholars pursuing two projects: to trace the historical "return" of environment to the center of biomedical theory and research, and to amend the concept of biopolitics to account for new forms and relations that redefine *bio* as medical and political in relation to different spatial scales and forms of life, such as oceans and microbes.[23] In addition, space biomedicine's exposed elite subject stands in political relation to other environmentally overdetermined and systemically endangered terrestrial subjects, such as those made through colonialism, disaster, and social vulnerability. Astronauts, with bodies damaged slowly by exposures to toxins and radiation, are made into high-profile and willing subjects of holistic environmental systems control.

## Risky Environments: 100 Kilometers beyond National Surface Space

Space biomedicine research aims to keep alive and healthy the people whom NASA's first medical director called "superselected citizens." These superselected citizens are spectacular emblems of biological selection, dramatically superseding those Sylvia Wynter describes as evolutionarily "dysselected" Darwinian human others.[24] They also constitute a small group of at-risk experimental biomedical subjects marked as national representatives being fit into an extensible whole: astronautically occupied outer space. Astronaut medical subjecthood is made possible by techniques that maintain life signs at the upper boundaries of biological survivability and governance. This celebrated at-risk environmental citizenship contrasts with the abject but nonetheless equally systemically emplaced subjectivities that come with technoscientific threats and disasters. In spaceflight, the threatening outer environment is an imposition on human biology rather than a provisioning space for it, and intervening in environmental support systems emerges as a distinctive form of "biomedicalization."[25]

My two-year internship with the National Space Biomedical Research Institute was in some ways an entrée to a familiar world of biomedical language and research, but NASA's biomedical landscape and discourse were unsettling. For one thing, there were restrictions on what I saw and where I went. I could not witness, for example, astronauts' clinical encounters. And I was inside a zone in which what was being secured from the public went beyond patient confidentiality or safety. In sites where I followed the production of the human system idea, I had to learn to think differently about what had been givens in my understanding of clinical practice. The categories of clinical practice and research resembled those in the earthly domains I knew well from my work in medical research project management before I became an anthropologist. But at JSC, "environment" never receded from view or discourse; instead, it modified everything I saw, heard, and even touched as an intern and ethnographer. In space, I was taught, most life activities, from eating to breathing, are complicated by microscalar concerns about human–environment interactions. Even the collection of samples for research is subject to space's disturbing effects on fluid flow, instruments and containers, cells, and molecules. Environment is a problem for conjoined bodily and technical systems in space, and astronauts have to embody solutions to it.

After people are selected to become NASA astronauts—a career marked by frequent and comprehensive medical and psychological screening—they are cared for as environmental test subjects as well as publicly emblematic civil servants. JSC's on-site Flight Medicine Clinic is at the heart of a complex constructed to showcase the fitness of the nation's sociopolitical system as well as that of its exemplary spacefarers. During and after their service in the corps, astronauts are encouraged to participate in longitudinal research. Among the buildings in JSC's cluster of cafeterias and program offices, the clinic faces a park and duck pond and is situated kitty-corner to the spaceflight program operations buildings and the Astronaut Office.

While in active service, each astronaut is routinely checked for, as one space biomedical clinician put it, any "otherwise insignificant problem" that might threaten the individual's space mission readiness as a "total performance system."[26] This phrase appears in sociologist Joseph Kotarba's analysis of space biomedical organization at JSC, in which he argues that aerospace clinical medicine takes an informally "holistic" approach to the "comprehensive" control of its "high-value" astronaut patients.[27] As does the military, NASA uses drugs to manage spaceflight human performance, but none of those I heard mentioned were geared to alter human physiology permanently. Kotarba describes how biomedical practitioners achieve control "holism" by assimilating the astronaut body and mind into totally systematized environments: institutional, social, technological, natural. Today, that totality spans a lifetime. A 2014 Institute of Medicine report describes astronaut lifetime health "surveillance" as an "ethical" response to the unusual and "uncertain" risk demanded of them, making work in environmental extremity a criterion for exceptional medical monitoring.[28]

Despite being inside protective suits and spacecraft, astronauts are exposed to conditions in outer space that make them temporarily or permanently impaired by terrestrial standards. After traveling past the 100-kilometer altitude that officially transforms them into "astronauts," people often experience forms of nausea and disorientation known as "space adaptation syndrome." Moving beyond Earth's atmosphere brings on a host of physiological effects, many of which parallel processes of bodily aging. With weightlessness comes fluid shifting that causes cardiovascular and hormonal irregularities and increased risk for renal stones, decreased red blood cell mass, and bone loss that may never be fully reversed. Although astronauts are more at risk of death from cardiovascular disease than from cancer, their endurance of high radiation levels still defines their occupational risk. Crew members wear dosimeters that track their lifetime space "exposure years," devices that heroically distinguish them from accidentally or disastrously irradiated subjects. However, the neurological and mental effects of spaceflight can alter astronauts in unheroic ways, making them act cognitively impaired or psychologically reactive when in orbit or back on Earth. When astronauts return to Earth's gravity, their bodies' adaptations to space, such as weakened skeletons, make them vulnerable to injuries.[29] To alleviate these problems, astronauts engage in pharmaceutical and exercise countermeasures aimed at protecting their "Earth-normal" health status while they also serve as the subjects of research on the effects of spaceflight on the human body. When I was conducting my fieldwork, research on these problems was a technical project referred to as "bioastronautics."

Ken's life sciences manager colleague Carla Prentiss, a white female psychologist with a past working in childhood development, described bioastronautics as "all things that address the human," from "medical care of the astronauts" to "the habitat [and] the environmental control systems." Thus, bioastronautics includes basic and applied life sciences. However, the funding for basic and applied space life sciences fluctuates according to presidential party agendas, leading some at NASA to speak of space biomedicine as "biopolitics."[30] At the time I began my research on the Bush-era Constellation program, funding ceased for basic biological research but became available for applied research into human systems as an ecosystemic element of larger mission systems. Practitioners' views of what constitutes that larger systemic space are broad

Space biomedical protocols prescribe "countermeasures" to replicate earthly environmental conditions, like weight-bearing exercise as a proxy for the terrestrial gravitational force that sustains "skeletal system" mass. Courtesy of NASA.

and often ecologically speculative, moving from the intermolecular to the heliospheric.

The ecosystemic integrations that mark astronauts as futuristic and fascinating human subjects include the astronauts' spectacularly intimate incorporations with technologies. The original speculative "cyborg in space" study imagined space–environment adaptation from within the body rather than through the use of spacesuits.[31] Contemporary astronauts, however, set limits to the kinds of accommodations they are willing to make in the name of adaptation, having objected early on to being "Spam in a can" or, in an only slightly more advantaged version of "subhuman" subjects that preceded them, "lever-pulling monkeys." The subhuman-to-spacefaring-human discourse is embedded in the evolutionary ordering practices of aerospace research—a process that retired NASA chimp trainer Joe Brady explained to me would continue as a future division of labor between earthbound human "senders" and those "sent" into space. Despite this evolutionary trajectory, astronauts today do not have machines routinely grafted into them, nor do they see themselves as an alternate species. And while the authors of the cyborg study imagined that astronauts would be spatially liberated by implant systems, astronauts live and work in a highly constrained environment. They actively participate in research to advance human–machine systems integration but also act to resist personal intrusions.

While I was assisting in the recruitment of aquanauts to participate as research subjects for NEEMO missions (see chapter 1), an aquanaut who was also a physician as-can reminded me that astronauts qualify as members of a "captive" and "at-risk" medical population who negotiate their personal bodily integrity. However, their position is actually more ambiguous than that. He and other astronauts I spoke and worked with expressed excitement about research in theory but also exhaustion and impatience with it. Because they are elite government employees participating in an extreme and experimental environment, they occupy a biopolitical limbo somewhere between privilege and abjectness. Although they are subject to intensive monitoring, social and medical surveillance, and demands to be researched, my interlocutors reported, as do other sources, that astronauts can refuse invasive procedures and monitoring they fear might lead to the detection of conditions that would disqualify them from flight.[32] One physician told me that during the

Apollo era, astronauts would joke that they "wouldn't be happy until the last physician is strangled by the entrails of the last psychologist." He illustrated this by describing an Apollo-era astronaut teasing a younger one wearing an experimental suit: "He was saying to the FNG [fucking new guy], 'Do you like looking like a medical experiment?'" There are also (as yet poorly) documented stories of uncompliant astronauts, such as the participants in the Skylab 4 "mutiny" in 1974, in which astronauts stopped working in protest against a grueling work and research schedule. Several flight controllers and physicians assured me that astronauts find ways to control their in-flight space and time, for example, by occasionally turning off vehicle cameras and microphones.

Even as astronauts negotiate their personal bodily boundaries and autonomy, it is ultimately their intimate relations with technologies that make humans into environmental systemic "elements" that constitute astronaut subjecthood. In this ecosystemic assemblage, astronaut bodies and life processes are not the only sites of biomedical monitoring, investigation, and intervention; those activities concentrate on the points where bodies and minds interface with technologies and environments. Such interfaces, from workstation setups and food/waste systems to spacecraft atmospheres, are also biomedicalized and monitored as systemic.

The "population" being systemically monitored in astronauts is, then, only partly *bio*political: it is a collectivity of living and mechanical things made to cofunction in mutually sustaining, if not biologically generative, association. NASA life scientists and engineers collaborate to design and manage semiclosed-loop environmental systems in which the "space module and its crewmembers are exposed to an intricate interplay" of "physical environments . . . , human responses to these environments, and the environmental design limits that are based on these human responses."[33] This work produces knowledge about human–environment interactions in order to make things compatible and to mitigate the "high probability of cross-contamination among crewmembers and between crewmembers and space module."[34]

In this model of human–environment interdependence, the definition of astronaut health is still based on biologically defined "natural" standards of wellness, but that definition must flex to include as "normalizable" those abnormal states that come about in the most extremely social and artificial milieus (e.g., spacesuits and marine suits and vehicles, artificial ecospheres) that social groups and nations have the power to create.

## *Going beyond Biopolitics in the Astronautical Environment*

As the human element in a national experiment to innovate ecosystemic interactions, astronaut medical subjecthood highlights the growing importance of environmental systems concepts in practices of modern governance. For this reason, it is important to consider how space biomedical control practices relate to other national projects concerned with remaking bodies and milieus simultaneously. As I discuss later when I locate space biomedicine in the contemporary genealogy of environmental medicine, the political-spatial concept of the organismal milieu is central to enterprises like colonial medicine, public health, eugenics, penal reform, and urban planning.[35] In such programs, sex, gender, race, and other socially and culturally defined characteristics become biopolitical targets for enhancement or erasure through biomedical management of bodies in space. U.S. spaceflight biomedicalizes individuals within built technical systems in order to extend human crews in general and Americans specifically into space.

Ironically, the life processes at the core of earthly biopolitics and space colonization imaginaries, such as sexuality and reproduction, are deemed unmanageable in governmental outer space.[36] As sociologists Monica Casper and Lisa Moore found in their study of NASA's designation of female bodies and identities as high-risk mission "factors," there is a persistent expectation for all crew members to exemplify restrained heterosexuality.[37] Two longtime life sciences technicians told me rumors of crew members having sex in space, but this topic was more often joked about as something for "private" spaceflight. In the early days of human spaceflight medical experimentation, the bias against sexually and reproductively marked female bodies meant that a group of white women passed early astronaut qualification tests but not one of them was selected to fly.[38] As a parallel to this form of bias and erasure, variations from an Anglo-American Christian "baseline" among crew members were explained to me by one psychologist as "cultural or behavioral" mission factors to be managed through "team cohesion" interventions or "accommodations" of diet and religious practice. Such sociocultural biases continue to happen even as astronaut selection criteria change. I heard social differences in the astronaut corps described in broader than official terms, as people spoke of "tree-hugger" and "evangelical pilot" astronauts and speculated about when the decades-long presence of gay astronauts would become public—which happened after the death of Sally Ride in 2012. In addition, the midcentury "right stuff" formula for selecting white military males has been replaced by other criteria associated with models of biological and ecological fitness. A veteran aeronautics psychologist told me, "We're now looking for adaptability and resilience."

Astronauts are biopolitical subjects in an ecological space, undergoing protected exposures to threatening conditions that are not conducive to lifetime monitoring in outlying spaces that are in precarious stages of social incorporation. In the following sections, I show how spaceflight's *bio*politics are not soluble in the politics of *eco*logy, but connect the two. To build toward introducing the concept of ecobiopolitics, I highlight how space biomedicine works with the human in ways that shift the "bio" off center in a systemic context in order to achieve spaceflight's extraordinarily scaled systemic relation building—from the molecular to the cosmic. I then describe ecobiopolitical subjecthood, which is a form of systemic life that goes beyond optimized biological health and fitness as well as normalized cradle-to-grave lifetimes.

## Biomedicine in a Solar Ecosystem

In its chartered mission to support terrestrial life outside its native environment, space biomedicine is one among many technical enhancement practices that advocate interventions to make human bodies more "flexible" by going "beyond what is 'normal' or necessary for life and well-being."[39] While the early U.S. space program emphasized the need for superhuman social and medical bodies, two of my biomedicine interlocutors reported that astronauts deviate from those standards of unstigmatized perfection today and that some have received formal "medical waivers" to fly. Other interlocutors observed that the development of better spaceflight environments will compensate for the limits of bodies and technology. For spaceflight practitioners today, the thing that extends "beyond" in space and time to enhance human life categorically is not an enhanced biological being. It is instead a controlled environmental system.

Paul, a junior JSC flight surgeon trained in occupational medicine, put his commitment to ecosystemic human enhancement in these terms as we sat in a small beige conference room within the institutional maze of JSC flight medicine offices:

> So my idea and my goal as a flight surgeon in support of astronauts is to get them into their best condition and to reach their human potential in whatever environment they might be, whether it be on Earth and the ocean or off the planet. That's my job. And so I have to ensure that the human is in good shape, but I also have to ensure that the environment is going to support the human, and allow them to reach their potential.

Paul and most of my space life sciences interlocutors are among the 3,200 members of the international Aerospace Medical Association (AsMA), which publishes the journal *Aviation, Space, and Environmental Medicine.* Space biomedicine's role as an environmental biomedical practice makes practitioners orient themselves as scientific and visionary experts on human environments.

Aerospace life sciences emerged out of mid-twentieth-century American and German military flight medicine and human factors science (space physicians are still known as flight surgeons). After NASA's 1960s era of small-crew programs ended and the larger space shuttle and station programs began, aerospace medicine became one arm of a life sciences program. Although in the United States space life sciences is a field constrained by turbulent changes in governmental funding and policy, the mission statement of the Space Life Sciences Directorate during the Constellation program—"to be the world's leader in understanding the space frontier and the opportunities, capabilities, and limitations of humans living and working on that frontier"—neatly expresses the field's persistent historical goal of remaking the scope and scale of human milieus.[40] The practices of aerospace medicine decenter and distribute the human body in ways that distinguish it from normative biomedicine.

Although it is an elite practice, space biomedicine belongs within the relatively marginalized practice of environmental medicine. This marginalization is related to modern biology's emphasis on the organism rather than its environment and on the practical authority of biology over medicine, as Canguilhem describes. By the late twentieth century, the compound term *biomedicine* signaled the triumph of biology over medicine as "the ultimate description and account of disease origins and mechanisms"; the term is also linked to the beginnings of government-sponsored environmental medical experimentation.[41] Historian Peter Keating and sociologist Alberto Cambrosio note that before 1970, biomedicine was specifically defined as "a discipline concerned with analyzing human tolerances to different environments and with methods of protecting against the effects of exceeding these tolerances" that was emerging in the aerospace and nuclear industries.[42] This suggests that for experts adhering to this definition of biomedicine, the prefix *bio-* authorizes medical practice to experiment with ways to manipulate human bodies to survive the environmental conditions to which new technologies subject them. Space biomedicine retains its commitment to both of these disciplinary antecedents: medicine as an environmental practice and medicine as an authoritative tactic for adjusting humans to environments.

Environmental medicine begins with Western medical control over the clinic's most vital outlying spaces: colonial territory and the interiors of private homes. Scholars who trace the rise of the pathogenic model over the "climatary" model of disease causation in the nineteenth century show how this process directed colonial medical spatial practices, such as tropical medicine and public health, toward managing threatening biological agents and bodies.[43] Vladimir Janković traces the emergence of environmental medicine to England and Victorian-era concerns with controlling interior "climates" to make moral bodies and minds, paralleling other projects aimed at disciplining populations through architecture and engineering.[44] With the genomics era, however, manipulating milieus became secondary to intensifying manipulations of biological material in and of itself. For these reasons, environmental medicine remains a small subspecialty of biomedicine with jurisdiction over an ambiguous class of environmentally induced illnesses. As such, it is linked to occupational and preventive health and continues to be described by advocates as a "missing element" in medical education.[45]

The AsMA website describes the association's members as "dedicated to enhancing health, promoting safety, and improving performance of individuals who work or travel in unusual environments" and defines those environments as "extending from the microenvironments of space to the increased pressures of undersea activities." It further notes, "Increased knowledge of these unique environments of 'Spaceship Earth' helps aerospace medicine professionals ensure participants are physically prepared, physiologically safe, and perform at the highest levels."[46] By implying that all kinds of human milieus, even nonnormative ones, can be made vitally commensurate, AsMA displays a vision of environmental systems scalarity. The spaceship-planet connects to ecological systems theory but also to a NASA ideological spin-off created by a self-described "planetary physician," chemist-physician-biophysicist James Lovelock. His now-famous "Gaia hypothesis" posits that humans are part of a cybernetic self-correcting and self-optimizing environment.[47] "Life," in AsMA's vision, is a matter of situated existence in interrelated microenvironments that are not natural or unnatural but, rather, simply usual or unusual.

The focus on charismatically "unusual" spaces like space or the sea, however, highlights how the biomedical bodies and sites of environmental medicine are sociopolitically partitioned. Space biomedicine makes research and care resources available for predominantly white and male bodies, mitigating their relationship to industrial environmental exposure. These protected bodies stand in stark contrast to other occupationally exposed bodies. This pushes biopolitical relations into an environmental political frame, where space biomedicine's selected "human" is arranged within the spaceflight cultural scalarity that situates the bodily system within a cosmic system.

## *Practicing Biomedicine in a Cosmos*

Space biomedicine practitioners claim specially situated knowledge as environmental medical experts, placing "the human species" within a cosmic technoecological context. In this perspective, environmental biomedicine and engineering are interrelated as strategies that produce evolutionary fitness. Although past and current NASA physicians and life scientists complain about working as a disciplinary minority in an organization "by and for engineers," they claim to be well versed in aerospace engineering and environmental science and technology. Most of the engineers and life scientists I interviewed agreed that one of NASA's virtues is its "big picture" view; when asked to describe what NASA produces, people often responded with descriptions of how they interconnect things and spaces. A senior planetary science manager with a background in engineering explained, "We [are] making the opportunity available to integrate the knowledge, to understand our solar system, our universe, and our place in it." A flight surgeon training to participate in a spaceflight analog excitedly described the ways future physician-astronauts would embody "integrated" knowledge:

> Because when you have astronauts who are physicians . . . this person's primary responsibility is to take care of other humans and human systems on this ship. And their secondary thing is that they're gonna be a navigator, they're gonna be a robot repairer, they're gonna be a geologist, they're gonna be all those things.

Interlocutors supported these arguments with references to a human "destiny" to colonize space, often in a pragmatic tone that effaced colonial histories on Earth. In this colonial model of spacefaring as a species destiny, a collective human future is shaped by interactions between exploring groups and the inhuman nonterrestrial ecologies they make and adapt to.[48] Both space biomedical and colonial imaginaries valorize the effects of extreme ecological adaptation (usually tropical) under technical control, rewriting the biopolitical goal of maintaining populations within a biologically normal state.

Ken's reference to the ideal of an "optimized" human system, noted earlier in this chapter, highlights how spaceflight experts consistently emphasize the dual goal of optimizing both life *and* milieu. Both engineers and life scientists commit to these optimizations in practical as well as visionary and ethical ways. I discussed this with Sam, a flight surgeon who had recently returned from participating as a crew member in NEEMO. Sam's enthusiasm for putting experimental ways of life "into operation" became evident as he described the "joyful" effects that extreme embodiment had on his scientific and ethical understanding of environmental integration:

> And as an aquanaut, what a better way to understand humans and our environment is to go to an environment which not only makes up most of our Earth, but also we came from, and we, if you think of humans, I mean we carry the ocean within us, we carry the saltwater within us, we carry the electrolytes within us, we carry an

> environment within us, we carry our gills within us, our lungs, and our water within us, and I balance out my understanding of creationism with my understanding of evolution, by saying what a beautiful creature we are.

When I asked Sam what NASA's new policy to "extend human presence into the solar system" meant to him, he pointed to the need to improve living modes in an anthropomorphized cosmos:

> We're continuing to discover things in the solar system that are going to help us here. So as a physician, and working at NASA, to enable us to extend our human civilization is going to help me in my own specialty, and what my own goal is. . . . We're going to be able to take humans and their environments where they would never have gone. . . . We should do it smartly to enable us to extend our lives and do good things while we're alive. As a spaceflight scientist, what is it, I mean, to be able to get an entirely new door, open up a door to another environment. . . . I think if we extend the human presence, the solar system understands itself.

Sam was not alone in claiming that the cosmos understands itself through human awareness—this was a sentiment that I heard often. It can be found in the script of Carl Sagan's *Cosmos* television series and, as other interlocutors informed me, in the work of Pierre Teilhard de Chardin, Julian Huxley, and the Passionist priest and "Earth scholar" Thomas Berry. Sam situated himself within a recursively optimizing human–milieu systems feedback process: "If we were able to push our environment and our biological limits by pushing equipment that supports that, then we'll be able to trod less on our [earthly] environment."

In a bustling coffee shop down the road from JSC, Sam's senior male flight surgeon colleague Kyle told me that he was "taught to think in systems" through his love of architect Buckminster Fuller's work. Having worked medical and research protocols for two decades, Kyle told me that he believed the space program could be a species lifesaver by making humans realize their environmental contingency and vital inborn capacity to adapt to, control, and remake milieus on a cosmic scale. Fuller's work inspired Kyle to "start from the universe," to "be a good physician," and to analyze problems systemically:

> Well, now . . . we're all in the same boat. We're all in this environment, we're all in this sea of systems. Interworking systems. And if you screw up this part of it, there are intended consequences and there are unintended consequences. . . . So, we're not separate from the universe.

For Kyle, understanding environments systemically meant that Earth should be called "ocean," "outer space" should be called "radiation," and the changing galactic position of Earth in the solar system "may be a cause of pandemics."[49]

Arguing that space exploration "defines our humanness, and if you retreat from that you're retreating from humanness," Kyle defined the environmentally situated human that space biomedicine works with as a clinical and political subject. Although Matthew Hersch argues that NASA aligned with environmentalism as a politically expedient public relations strategy, interlocutors I spoke with described professional and personal commitments to improving human–environmental relations.[50] Spaceflight and medicine's political milieu, then, is explicitly environmental.

## Ecobiopolitics

As is clear by now, for spaceflight's life science practitioners and policy makers, a biocentric perspective is untenable; instead, what matters medically and politically is the capacity to know and manage vital milieus. With this in mind, I modify Rabinow and Rose's definition of Foucault's biopolitics concept to specify "ecobiopolitics" as a process centered on life–environment relations rather than on life itself: *truth claims based on knowledge of life–environment relations, power relations that take life–environment relations as their object, and the modes of subjecthood and subjectification that designate subjects as elements of environmental relations.* Despite its coming out of a fieldwork project on outer space, ecobiopolitics is not a far-out concept. The growing tendencies of modern societies to protect the calculably vulnerable "vital systems" supporting "collective life," such as energy systems and financial systems, situate these vital systems within territorial spaces like nations and homes as well as in intangibly bounded spaces like environments and ecosystems.[51] And in increasingly powerful regimes of environmental citizenship and ecological governance, biomedical and environmental governance are being sutured together, not collapsed into one another.[52] As I continue to argue throughout this book, system is the relational technology that does the suturing.

## *Ecobiopolitical Strategies: Environmental Normalization and Systems Risk Management*

When JSC life scientist Ken asked me if I was aware of NASA's "space normal work," the topic had come up because we were talking about new long-duration spaceflight technologies, such as the "Robonaut" designed to work outside spacecraft or on planetary surfaces. Gendered male, Robonaut has a humanoid head and torso but only one lower appendage, which is designed as a grappling hook, so "he" must sit in a wheelchair on Earth. The robotic astronaut is meant to work on the outside of spacecraft, virtually controlled by a human astronaut inside. "His" designers describe "him" as the ultimate "space-adapted" human body—legs are fairly useless in space. I mentioned to Ken how this highlights the social irony that astronauts launch into space as exemplary subjects but become automatically impaired or disabled once there. This is when he told me of NASA's efforts to define a "space normal" physiology. This is one of two ecobiopolitical strategies I describe here: one that dually normalizes extraterrestrial bodies and spaces, and another that treats risk and risk-mitigating interventions as intersystemic.

One of the hallmarks of biomedicine is how people and ways of life are marked according to the statistical binaries of normal/abnormal or normal/pathological.[53] Even as contemporary biomedical experts modify definitions of normality and pathology, these definitions must be coordinated across disciplines and spaces. Space biomedical practitioners, like their Earth-focused counterparts, are also intent on defining "space normal" statistical values for bodies, but doing so raises the problem that outer space conditions are, in terrestrial terms, abnormal—which is a problem for governmental regimes of spatial control that operate through normalization.

NASA's "bioastronautics risk reduction strategy" focuses on normalizing bodily responses to space but also has to account for the extreme bodily unboundedness of space living. In technically austere language that stands in striking contrast to Sam's and Kyle's more romantic visions of the human in space as a transcendentally evolving subject, this risk reduction strategy posited by life scientists working to make biological functionality legible to engineers defines the human as a "critical system of space flight" that has "operating bands" (i.e., upper and lower performance limit ranges) that "must be understood, controlled, and specified, as well as optimally integrated with other systems."[54] While such quantitative standards help establish stable parameters for spacecraft size and environmental conditions like air quality as well as how to keep astronauts "fit" enough to keep their machines working, experts I spoke with self-consciously worried that "space normal" body values can never be stable like earthly values. All of my interlocutors expressed some ambivalence about whether "space normal" or "space adaptation" can actually be defined as a natural physiological state, since it is so contextually and conditionally labile. This problem is echoed by Canguilhem's argument that normality is not a fixed value but a normalizing capacity, spurred by pathologies, to adapt to environmental variation. As Ken told me, "normal" is always "confounded" by the "myriad" artificial "microenvironments overlaid . . . on the human experience of spaceflight."

As a result, I found an actively reflexive recognition among space scientists and engineers that Earth and space environments are scientifically, socially, and politically coconstituted. Social theories that emphasize the equally scientific and political construction of ideas like "environmental adaptation" would not necessarily be alien to space life scientists. Ken and Kyle both complained that "space normal" is a just way of describing "medical countermeasures normal," meaning that it is an artifact of drugs and exercise rather than pure bodily adaptation. Kyle also offered that any idea of astronauts as "Earth normal" is inaccurate because it refers only to "normal with respect to subpopulations on Earth that meet these select-in criteria, that don't meet the select-out criteria." The idea of "Earth normal" is problematic as well. A group of NASA life sciences researchers have declared that spacecraft environmental design can never actually reproduce "Earth normal values" but can only find the most efficient ways to reconcile the socially determined "physiological, engineering, operational cost, and safety considerations" that constrain spacecraft environmental design.[55] The patently social conditions of "normality" that spaceflight practitioners acknowledge also captured the attention of disability scholar Victor Finkelstein, who argued for recognition that disability is socially defined by using the example that an astronaut's accidental disability or death would never be considered to "reside in himself" but would be seen as the failure of socially produced technologies and life-support practices.[56] Ken summed all this up for me: "Spaceflight is the ultimate social activity . . . nobody survives there by themselves."

The space biomedical recognition of normality as environmentally and socially determined echoes other astronautics practitioners' evolutionist theories of how human bodily optimization is intimately tied to social capacities for environmental optimization. A Canadian white male extreme sports enthusiast and physician turned astronaut, John, echoed this perspective as we talked outside the JSC cafeteria one day. He and other interlocutors claimed that the desire for "optimization" is a part of human nature, coded into human beings on a molecular level. As John put it, "Everyone has this little 'exploration gene,' if you want to call it that." He did not refer to controversial eugenics-style research on "proclivities" for "novelty seeking"; rather, he claimed that this "little gene" expresses "itself" as a capacity to transcend biological limitations and to "optimize" living spaces.[57] This theory is rife with the situated Western vitalism that informs Canguilhem's universalistic claim that any "physiological constant" is the "expression of a physiological optimum" that "the living being[s], and *Homo faber* in particular, give themselves" and that "flourishes in man in the form of technical plasticity and a desire to dominate the environment."[58] John's "everyone" has scalarity:

> The human species are extending our capabilities, we are reaching out into our own solar system, to extend our capability to live elsewhere in our solar system. And then putting that back in the context of theories of evolution and human origins here on Earth . . . the proposal I guess is that biology allowed humans to adapt and evolve to optimize their capability to live here on Earth, [and now] it's really technology, not biology, that's enabling us to live successfully in these environments.

In space biomedicine, then, the signature aim of biopolitics—to define normal life in terms of biological norms and vice versa—is not the most vital work to be done.[59]

Instead, the trick is to manage the technical adjustments of humans and environments; therefore, converting astronauts from de facto patients to "normal" subjects is tied up with normalizing outer space as a vital national and global environment. Therefore, the ecobiopolitical goal of "space normality" is to facilitate an intrinsic "human" drive to adapt. If, as Foucault notes, the modern clinic was born as a space to manage health as a governmental politics of life itself, space biomedicine naturalizes a politics of ecosystemic normalization and optimization. In the next section, I show how space biomedical practitioners deploy ecobiopolitical techniques to make living, mechanical, and environmental systems and their risks measurably interdependent.

## *The Human in the Loop: Probabilistic Risk Management*

During our meetings, Ken gave me more examples of "paper NASA": a set of digital illustrations that juxtapose space shuttle systems with their human system equivalents. A gendered "invisible man" merges with machine schematics in a series that moves from surfaces to revealed insides. The vehicle exterior corresponds to the integumentary system, the engineered structure to the skeletal system, the electrical system to the neurological system, and the power system to the metabolic system. Ken and his colleagues intended these graphics to be used in NASA outreach activities to engage American publics. As illustrations of astronautics as an integrated knowledge domain, the analogies in the images create a sense of systematicity. By portraying human and mechanical systems as complementarily vital and mutually vulnerable mission systems, the illustrations exemplify spaceflight strategies for control of what NASA scientists and engineers call "integrated risk environments." Like other biomedical subjects, the spacefarer is made ontologically "multiple" through distinctive expert practices that are both institutionally divided and coordinated, but in this case practitioners coordinate around the risk management of bodily and machinic systems.[60]

Risk is ambiguous at NASA. It presents a conundrum for practitioners who dedicate their lives to minimizing risk but who also celebrate it as generative for national and species development. Ken, like most of my interlocutors, spoke of childhood engagements with space things, places, and ways of being: drawing spacecraft, watching space missions, reading science fiction, building spacecraft models, and learning to do

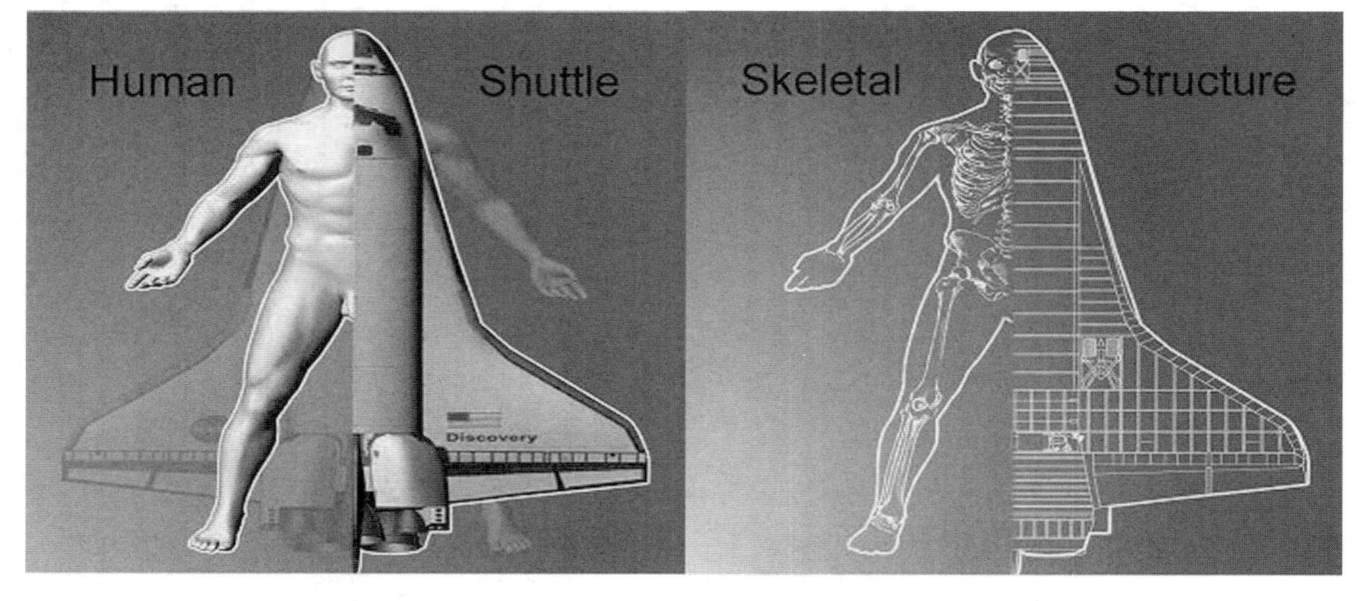


A NASA Life Sciences Division–commissioned graphic of analogized human/machine systems normalizes their interactive systemic natures in relation to the space environment. Courtesy of NASA.

risky things in challenging environments, such as diving. As in the case of John, many began these activities at as young an age as possible. Such engagements were, for these people, not just practice for future jobs but acts of staking out a socially risky alliance with "geeky" things and people. Testimonials posted by men and women in equal measure in the "When Did You Know?" section of the American Institute of Aeronautics and Astronautics website recount being called to space technology work.[61] These narratives have unacknowledged parallels with "coming out" practices and other ethical discourses about learned identities. Like his colleagues, Ken wanted to be an astronaut "as far back as I can recall," linking childhood play and his career decisions:

> [On] a sandy dusty playground on a little cement bridge . . . I'd lay on my back and put my hands up on the handrails and pretend I was John Glenn. . . . [I] tailored my education to become relevant to the space business . . . to demonstrate to NASA I knew how to do complex integrated things.

Even interlocutors who were not interested in becoming astronauts reported seeking risky and integral embodied experiences, such as participating in space analogs, flying on "vomit comet" parabolic aircraft flights that simulate weightlessness, and doing extreme activities (de rigueur on astronaut résumés) such as flying, rock climbing, and scuba diving. Although Ken described his job as "retiring" risk, he and others voiced disappointment in what they saw as a tendency toward risk averseness in U.S. government and society and, as Ken said, a disappointing lowering of the space program's "risk threshold."

People I encountered and worked with across NASA sites often described risk taking as an "innate" human characteristic. Nonetheless, among these professionals, the prime task of normalizing the integration of living and nonliving mission elements requires a new way of tracking risk that decenters the human as a special mission element and makes human life processes systemic beyond the body. Toward the end of my fieldwork, I heard that space life scientists had begun to put their "space normal" clinical and epidemiological data to work as risk management data aimed at improving the systems integration of Constellation's Orion crew exploration vehicle. Using a technique called probabilistic risk assessment (PRA), borrowed from the domains of nuclear power and transportation risk management, space life scientists advocated merging human medical data with hardware and software data in engineers' overall calculations of space mission risks. This strategy makes new truth claims about the nature and proper methods for systems integration. It makes astronauts into data-producing components whose likelihood of functionality or failure, otherwise known as health or illness, can be statistically represented as connected to other, nonliving mission factors. As life scientists often emphasized, their position as experts on "squishy" things in an engineering practice environment means that they must, in one life-support technician's words, "give the numbers" (i.e., quantitative data about health and safety risks) to engineers in order to manage using "hard data" on what human factors engineers call the "human in the loop."

By using PRA, space life scientists are attempting to convince engineers that humans do more than occupy vehicles or act like—in the words of one environmental systems flight controller I interviewed—"wild card[s] at play" in an otherwise controllable engineered environment. Life scientists present astronauts as part of a system where risks to environmental habitability matter for beings *as well as* for hardware and missions. In an article on the potential uses of PRA, space biomedical researchers note that NASA defines risk as "the combination of the probability that a program will experience an undesired event and the consequences, impact, or severity of the undesired event, were it to occur."[62]

Using the logic of environmental medicine and public health in an ecobiopolitical way that reverses the direction of biopolitical effects, these life sciences authors argue that the best astronaut-health risk intervention may actually occur in a mechanical or environmental system and that the best system-wide risk intervention might actually begin in an astronaut's body rather than outside it. In this model, human and nonhuman systems exist in an intimate milieu in which they are mutually at risk and present risks to one another. I spoke with a researcher developing the PRA project, Jane, a young white neurobiologist with self-described history as a southern debutante and an interest in quantum and systems biology. She explained the imperative for redefining mission systems, including "the human system," as contiguous rather than discrete. At first, Jane and her colleagues wondered, "Can we even apply [PRA] to the human?" But they found that if they could make "the squishy nature of the human system" calculable, this would make it easier to "break it all down" to engineers about why humans "affect your vehicle and how your vehicle can affect [them]." Jane, like a crew nutrition specialist I interviewed in JSC's food lab, spoke of "food systems" in the mission loop:

> So if [astronauts] have an inadequate food system, how does it break down. . . . Well, . . . it could be contamination, it could be the food doesn't taste good, it could be there's not enough vehicle resources, that's where the vehicle starts coming in. . . . You get into the food's unsafe, well maybe the food's unsafe because of something going wrong with the vehicle like the storage isn't right, or something . . . and so this is where it leads toward the human, now if the human gets sick, they can't do maintenance, they can't do some of the things you want them to do. And so we're starting to put the human in as another subsystem within the vehicular system.

To help me understand the connections between their space normal work and their PRA work, Ken, Jane, and others gave me documents that use biological descriptors for machines (e.g., "PRA is a systematic and comprehensive methodology to evaluate risks associated with every life-cycle aspect of a complex engineered technological entity")[63] and technical descriptors for human beings, such as "mission elements."

In these writings, humans, technologies, and their environments overlap and merge in an open and lively exchange of systemic materials and forces. These documents of "paper NASA" explicitly treat humans as mission environment "tools" rather than as operators. Others, as Matthew Hersch describes, treat nonliving but vital mission "elements" like checklists as humanlike actors or "crewmembers."[64] Although PRA calculations and spaceflight human/nonhuman categorical slippages seem esoteric, they formalize an environmental system's vital human–nonhuman associations and connections.

## Connected beyond Life and Death: Systemic Reintegration

"I can't say anything about that," the young Latina risk management lead technician said, shaking her head. She fell quiet, and I nodded, looking down. We could hear the sounds of employees talking elsewhere as we sat silently in a conference room in a corporate JSC contractor building in downtown Houston. Three years after the space shuttle *Columbia* disintegrated during reentry in 2003 when damaged thermal tiles failed and the vehicle caught fire, I interviewed this contractor, who coordinated a team to recover debris scattered across miles of land. Our exchange was awkward and inflected by mutual anxiety about discussing the crew's bodies, how parts were found, collected, treated. She gave refusals, incomplete sentences, hints, and deflections to my questions about body recovery—all part of the accumulating but often publicly hidden domain of spaceflight story and experience.

Like others I met who knew the STS (Space Transportation System) 107 crew as coworkers and family members, this contractor gave me ways to perceive the connections being made between astronautical life and death, physical and symbolic. One day in 2008 when I watched two shuttle launches broadcast live from Cape Canaveral with JSC employees and families on a big screen in the JSC auditorium, I sensed the contiguity of life and death around me in the noncontradictory display of tears and elation. Some spaceflight practices of memorialization, like this service for a JSC employee in which the deceased was spoken of as being on a journey through the solar system to heaven, can be understood in standard anthropological terms as forms of social-spatial integration. Commercial launch companies like Celestis now send the cremains of any paying customer into orbit on a "memorial spaceflight" or on a "journey" into the solar system, thereby creating a "posthumous cremain-astronaut."[65] In spaceflight, such practices fit within a specifically ecosystemic cosmology, along with other U.S. institutional practices that systematize the relational conditions of life and death and beyond.

In writings aimed at a popular audience, U.S. systems dynamicist Donella Meadows describes the dead body as an example of something that "loses its system-ness"—until it becomes part of other systems.[66] Meadows coauthored the controversial 1972 book *Limits to Growth,* a report commissioned by the Club of Rome that modeled an impending clash between economic growth and the finitude of the Earth system. Meadows's statement describes a moment of modern processual disjuncture—not between life and death but between systematicity and nonsystematicity. Thinking of thought and life in a modern systemic processual mode, Foucault situated his theorization of biopolitics within his discussion of extant and competing historical "systems of thought," which implies that his parallel concept of the twentieth-century thanatopolitical administration of death is the product of equally systemic processes.[67] But the fallen astronaut, a spectacularly revered public figure, is not the kind of thanatopolitical subject that theorists can describe as dispossessed of legal political systemic life.[68] The deceased astronaut as an institutional subject occupies a theoretical place closer to that of the "twice dead" brain-dead patients that anthropologist Margaret Lock examines, who are put on life-support systems to exist in a technically defined and politically arbitrated slot on a life–death continuum.[69]

Just as biomedical life-support systems make lifelike forms of death, the system concept, as a relational technology of space and time, can keep living beings systemic beyond their biological ends. Death, therefore, is not necessarily a systemic end, as ecologists argue when they map out the recycling of matter and energy in ecosystems. In this way, even the cruel neoliberal condition of actuarially consigned death that philosopher Eric Cazdyn describes as constituting populations of the "already dead" can be articulated as producing economic systemic vitality.[70] Death in ecobiopolitical terms can be ecologically distributed, so that individual bodily deaths and species extinctions can all be incorporated as parts of larger manageable ecosystemic processes.

But astronaut death is also a social systemic failure at scale. The *Columbia* accident, a loss intensely experienced within the agency, led to acts of managing systemic fragmentation, gathering in and making whole again. While human death can be abstracted to be like the failure of damaged hardware, it also represents a failure of agency and government to manage what interlocutors call the "mission environment." Exasperated physicians told me that crew escape technologies and even body bags were controversial elements of spaceflight systems because they materialized the possibility of failure. So, as *Columbia* resulted in public memorializations of crew members, it also generated internal discussions and investigations of failure points in systemic processes, such as the "normalized" deviations from standard procedures that sociologist Diane Vaughan identified as a cause of the 1986 *Challenger* disaster.[71]

A 2014 NASA final "aeromedical lessons" report on the *Columbia* accident, titled *Loss of Signal,* identifies the disaster as a "multisystem" failure—but also gathers those things back into other systems.[72] The report begins by describing new "collection and identification" systems for recovering bodily remains, sorting field-recovered objects, and reconstructing the craft. It recommends rethinking the crew's bodily equipment, from helmets to restraint belts, as an "integrated system" with evaluable "weak points" of connection with other spacecraft systems.[73] It ends with a section authored by Jonathan Clark, the physician husband of *Columbia* astronaut Laurel Clark, that presses eloquently and passionately for better "survivability" systems than escape processes.[74] Such postaccident reports emerging from shared grief offer suggestions for technical responses in an attempt to reassemble spaceflight itself as something systemically vital rather than terminally risky.

In life and death, U.S. astronauts are environmentally systemic beings who make national systematicity more than terrestrial. In 2014,

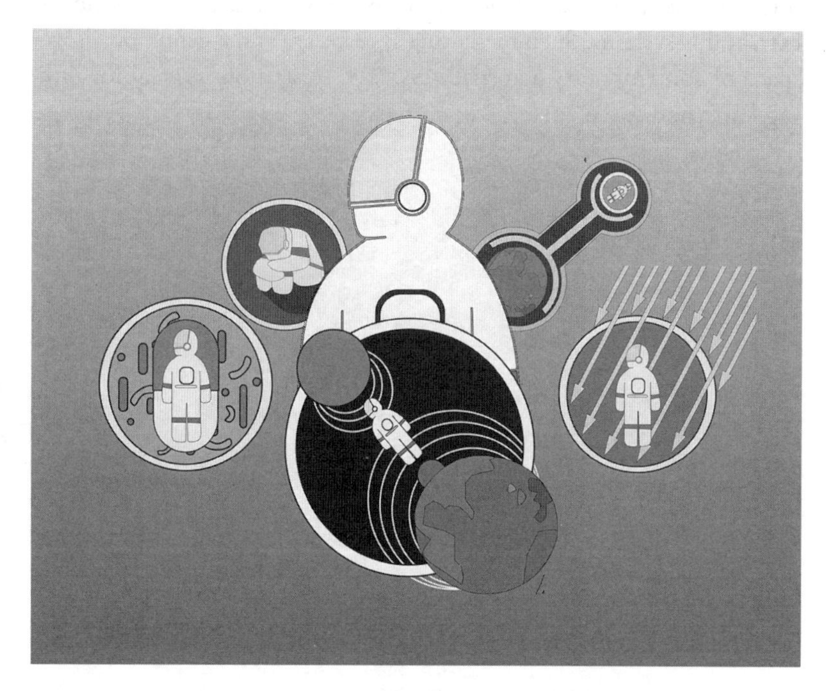

NASA Human Research Program logos for its "Mars forward" environmental health research foci (clockwise from left): gravity fields, isolation/confinement, hostile/closed environments, radiation, and distance from Earth. Courtesy of NASA.

NASA began conducting its Twins Study, coordinating twelve university-based investigations to "compare the molecular profiles" of identical twin astronauts Scott and Mark Kelly, "two individuals who have the same genetics, but are in different environments for one year."[75] Calling this "exploring space through you," the study website analogizes its microscopic investigation with the development of better space-based telescopy.[76]

As exemplary ecobiopolitical subjects mediating the connection between micro and macro spaces, astronauts live out the social systematicity and scalarity of environmental knowledge production and management. Such systems practices are ordering modes characterized by control of both connectivity and separability. In the next chapter, I examine the conceptual and material transformation of an iconic "separation technology": the spacesuit. Just as life scientists work to fit bodies into systems, they also participate in making spacesuits into kinds of extrabodily systems. Such mediating systems are at the center of a larger space of contemporary aerospace systemic separations crucial for linking Earth and space.

# 3

# SEPARATION
## BUILDING PARTIAL BOUNDARIES

Building 7 is like a museum of systems. It contains the Crew Systems Laboratory, formerly known as the Life Systems Laboratory. The building's 1960s-built foyer and halls are lined with display cases full of changing kinds of body-shaped environmental boundaries. Spacesuits, gloves, and helmets from the agency's programmatic history stand and sit in dioramic arrangements according to their eras and functions, like taxonomically ordered technical forms. Their differences exhibit alterations in the very shape of "life support" as a construct with intimate and exteriorized dimensions. In summer 2007, I walked the building's shiny vinyl-tiled floors with an experienced and ready-to-retire biomedical engineer, Rick, a white male born and educated in Indiana and hired at JSC after the Apollo program ended in 1975. The halls we passed through were redolent with the scent of old-school building materials and hypercool vented air that separated us from the stewing Texas heat outside. Rick was orienting me to the display, case by case, once in a while extending a hand, palm open, almost lovingly, toward the glass dividing us from this humanoid equipment. The equipment's systemically integral parts were designed to provide a terrestrial body with the exterior coverage, atmospheric pressure, thermal regulation, and circulating air needed to withstand the dangers of outer space.

We stopped in front of an Apollo-era spacesuit, an assemblage of flexible dull-white fabric and heavy hardware crafted into an inflatable bodily envelope. Before telling me which company built it (I guessed the David Clark Company—its motto: "Setting the standard of excellence for air/space crew protection"), Rick pointed to shiny vent fittings meant for hoses leading to a portable life-support system backpack to manage the circulation of air and breath. "Look, " he said, "the port colors correspond to arterial and venous blood flows so people won't forget which coupling is which." Mindful of what it would be like to wear that suit, I observed, "When it's full of air, it seems like it would feel so detached from the body." He countered quickly, giving that sense of separation a kind of feedback function, "I've heard crew members say that the sound of whooshing air in there comforts them. Means it's working." A week earlier, when I asked Rick about his perceptions of what it might be like to be in space, he leaned back in his office chair and looked up at the ceiling. He exhaled slowly, then drew out a reply:

> When I think of space I think of a vacuum. I get a cold feeling in the pit of my stomach. When you say space I think . . . not about life or absence of life, but the challenge to our life that being in space brings. . . . Scott Carpenter [Mercury astronaut] said space was quiet.

People I interviewed often mentioned mediated senses of being in space—the quiet of the early capsules, the grating noise of the space station, the scent of gunpowder reported by "last man on the moon" Gene Cernan. Rick's "whooshing air" is the good sound of death-dealing conditions kept at bay, in a suit built to keep life in but apart from space. His biomechanical systems sensibility echoed those of other engineers I knew who expressed the value of making such vital intimate separations possible. Today at NASA, environmental engineering design and redesign work is understood as inherently historical and is explicitly recognized as a cultural resource. When I was doing fieldwork, Building 7 was being evaluated for inclusion on the National Register of Historic Places. Although it has yet to be formally added to the register, it has been recognized as a pivotal JSC site—like the Space Environment Laboratory and the historic mission control center—in which national environmental expansion history is being made.

Houston, Texas, is full of protective suit "systems" designed to selectively connect and separate bodies and environments. A longish walk

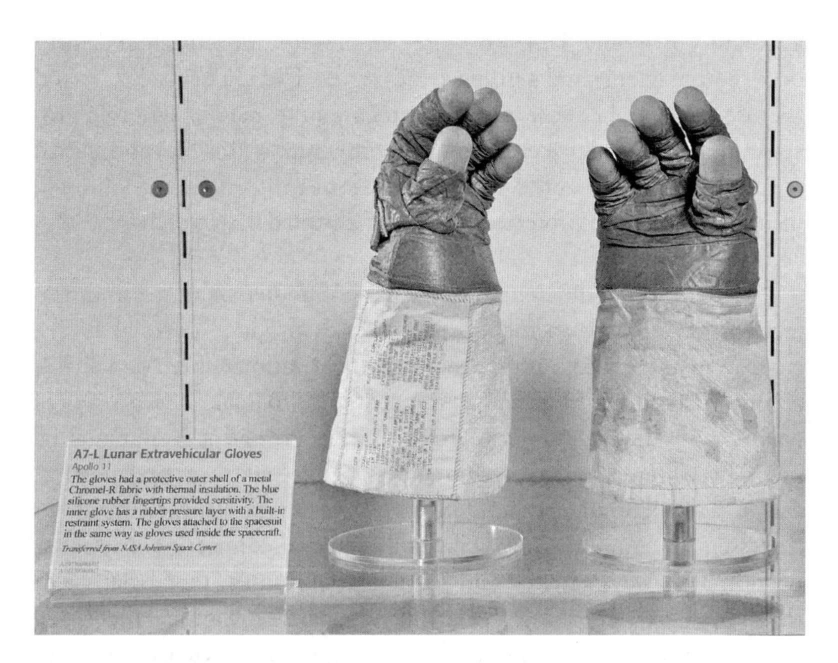


**After the death of astronaut Neil Armstrong in 2012, the National Air and Space Museum put his helmet and gloves on limited display as iconic elements of lunar life-support systems requiring special forms of preservation against deterioration. Courtesy of NASA.**

from JSC's Building 7 leads to the Lunar Sample Laboratory Facility, where I watched people in bright white cleansuits and layered gloves enforcing "planetary protection" protocols for maintaining the biological separateness of planets and uncontaminated machine–planet connections. As I went out to do fieldwork within Houston and in the broader Texas landscape, I found scattered spacesuit wholes, bits, and parts being used in museums and classrooms to demonstrate the potentials of controlled human–environment interactions, and I saw a famous female space shuttle commander's spacesuit displayed in a private home collection. I went with JSC physicians to watch the Red Bull "space dive" science team prepare a suited Felix Baumgartner for an altitude chamber test in San Antonio. And in downtown Houston, petrochemical companies manage workers in scuba gear and hazmat suits to protect them from toxic materials. That work, too, has a spaceflight connection. People involved in designing and maintaining protective suits, including astronauts as mobile U.S. workforce employees, move back and forth between the energy and aerospace industries. The use of spacesuits and the use of petrochemical industry suits are both part of systemic processes that also produce everyday environmental toxins. Beyond industry production safety zones, those toxins are distributed outward through machines into the unprotected spaces and unsuited bodies of living things across the globe.

Even though differential treatments of bodily exposures and protections reveal how sociopolitical processes produce unequal body–environment relations, anthropological theorizations of relationality as a concept tend, as Marilyn Strathern argues, to make the process of connectivity their ontological ground.[1] That ground is legitimated by scientific claims about the essentialness of biological interrelatedness, even if, as both Strathern and Donna Haraway caution, any claims about the essential nature or universality of relational connection *or* disconnection are always situated and partial.[2] Importantly, this social scientific preference for making connectivity the basis of relational theorizing has crucial analytic and political value. It generates vital critiques of Western denials about social and biological interconnectivity and reveals the ideological strategies of social separation regimes, such as segregation and categorization practices, valorizations of individualism, and territorial boundary maintenance. Yet, following Strathern's and Haraway's observations, it is possible to see how such analytic perspectives can cast connection as generatively relational and separation as destructively antirelational. Morten Pedersen argues that such perspectives can "fetishize" cultural practices of connectivity and foreclose examinations of "creative" dividing activities that produce differences in relational ontologies that are not necessarily antirelational. He uses the examples of shamanic woodcutting and other "detachment" practices to illustrate his claims that disconnectivity is vitally constitutive of social relations.[3]

As environmental social scientists examine spatial relations in which connectivity is a threat, it is becoming clear how contemporary social life is shaped by the politics and techniques of variable body–environment connectivity *and* separability. Access to clean water and air can be as much a result of the politics of location as a matter of the politics of access to filtration and purification technologies that keep toxins separate from bodies. Today's landscapes of variable environmental separability result in, among other things, a process that Kim Fortun calls the "toxic vitalism" of unchecked contamination that nurtures capitalization and industrial development.[4] Fortun critiques theories of networked relations that confer a generative vitality on all interconnected things, even toxins. In doing so, she calls attention to how both industrialization and pro-network theorizations gloss over ways that differential hierarchies of unprotected toxic exposure are produced "within a given system."[5] Although she uses "system" as a descriptor for a given set of relations, her critique exposes systematicity itself as a political technology of relational connection and separation. Fortun and other environmental social scientists go beyond the connective politics of relationality to document the results of processes that keep some bodies and spaces *inseparable* from industrialization's connective pipelines, networks, infrastructures, and systems. Their environmental fields include toxic towns, racialized and class-based landscapes of pollution and allergens, and sacrificial wastelands that constitute broadly unequal spatializations of in/separability.[6] All in all, emerging environmental anthropological work shows how spatial relational ordering regimes rely on distributed coproductions of *both* connection and separation, much of which is enacted technologically.[7]

This chapter attends to how contemporary relational ordering powers and processes produce *separation technologies* built specifically to systemically separate things and spaces, and to do so selectively. Complementing the preceding chapter's focus on connection, this one centers on systems as relational technologies of separation. In what follows, I link together three aerospace separation technologies that empower environmental control on a solar systemic scale: spacesuits, "planetary protection" processes to control the boundaries between the terrestrial and the extraterrestrial, and space center environmental cleanup mechanisms. Analysts have examined the combined semiotic and political power of separation tools and structures, from sieves and stethoscopes to dams and border walls, but I focus here on spaceflight systems technologies specifically designed to control the interrelation of bodily and worldly environments.[8]

## Building Systems Out of Partial Separations

During Constellation, which was intended to build an environmentally connective systems architecture like no other, NASA administrator

Michael Griffin delivered a reflective lecture titled "System Engineering and the 'Two Cultures' of Engineering," in which he declared that systems engineering, unlike other forms of engineering, has a harder job to do than simply connecting things: "I like to think of system engineering as being fundamentally concerned with minimizing, in a complex artifact, unintended interactions between elements desired to be separate."[9] The agentially unattached term "desired" opens up questions about who authorizes the production of systemic separation technologies, how they connect technical and political intentions, and who and what are included in and excluded from those desires. The systemic separation technologies I put together here—spacesuits, planetary protection, and environmental cleanup—are crucial to the authoritative production of human–environment relations and scalarity. And not as solidly bounded separations, but as selectively membranous and intersystemic.

Following theorists of partial connections and Griffin's statement, I hold that separation technologies are powerful because of their capacities to separate *partially.* By emphasizing the intended partiality of these astronautical separation technologies, I attend to how contemporary systems work with parts is not just about making wholes and totalities. People also aim, in ways that formally and informally align with theories about systemic complexity, to manage things and processes understood to be semiclosed, partially open, and parts of multiple systems. And systems-minded scientists and engineers separate with partiality in the dual sense—as in Griffin's quote—of partitioning as well as of enacting preferential desires. Partial connections and separations are often not unconscious results or shortcomings of environmental systems thought and practices, they are keys to the purposeful legitimacy of such thought and practices. Therefore, I also focus on the aspirational, moral, and associative dimensions of spaceflight separation technologies. Just as the pressurized spacesuit generates a lively whoosh of air that signals the managed interreliance of crews, technologies, and technicians in the spaceflight enterprise, separation technologies, no matter how ephemeral or seemingly inert, actively work to control how and where disconnection operates as a relation-*building* process.

Which brings us back to spacesuits. The spacesuit is more than a suit; it is an iconic spatial relation-building technology that mediates nation-building labor. In a NASA educational publication for K–12 students, the introduction describes all human beings as "astronauts" traveling through space on a planet and details how spacesuits create portable environments that enable "work" on vehicles and other planets.[10] As working tools, they become comparable to other terrestrial labor separation technologies, reflecting the production of knowledge about what is required to separate particular bodies from specific forms of environmental threat and enable them to survive.

In the next section of this chapter, I follow how contemporary spacesuits are being redesigned to function not as total boundaries but as "extrabodily systems" that partially separate bodies and outer space. After that, I examine how protective techniques to keep Earth and space biologically separate act also to partially separate the distinct goals of spaceflight science and engineering and to keep technical problems of environmental boundary maintenance conceptually separable from problems of environmental ethics. Finally, I examine NASA's ongoing attention to terrestrial contamination around its facilities. Jet Propulsion Laboratory (JPL) activities concerned with sorting out the agency's responsibility for water contamination in the Los Angeles area provide a window into how NASA working groups enact selective separations between the agency's civil and military institutional identities and between its terrestrial and extraterrestrial environmental systems work.

## Spacesuits as Extrabodily Systems: From Protective Ships to Interactive Skins

The complex challenge of suiting up astronauts to live in outer space uniquely exposes scientific, technical, and political questions about selectively controlling human/environment boundaries and limits. The suited astronaut's mission to survive, move, communicate, and work in nonearthly spaces is currently enabled—albeit awkwardly—by the International Space Station's impressive extravehicular suit, known as an EMU (extravehicular mobility unit). An astronaut suited in the EMU's so-called "anthropomorphic independence unit" is a relational configuration that renders extraterrestrial environment workable. The U.S. suit/human unit's theoretical mission is to work out problems in places where life and milieu both pose difficulties for NASA's capacity to fulfill presidentially articulated national needs and desires and congressional Space Act intentions. In this way the suited astronaut is an aspirational ecobiopolitical intervention into life–environment interactions. The spacesuit operates as a separation technology on these levels but also, in the spaceflight workplace, has to function as a system that allows relations with other systems and spaces. While the EMU perpetuates the classic space-age idea of an environmentally hard-bounded system, during the Constellation program spacesuits were being tested even more rigorously as flexibly bounded extrabodily systems. They were also being rethought, in the "paper NASA" tradition, as new conceptual spacesuit system designs. In both activities, newer spacesuits displayed ambitions for more partial, rather than more total, environmental and social separations.

The pressurized extravehicular spacesuit is a spaceflight technical "icon," a status that inspired several cultural historical analyses of spacesuits that were published while I was in the field. If the Apollo-era astronaut is the most iconic of astronauts, the Apollo-era lunar suit is the spacesuit's most iconic form.[11] Analyses of that iconic form emphasize how the suit's manufacture ended up reflecting the needs of soft organismic bodies rather than gendered space-age visions of the steely, hard, "right stuff" symbolic masculine bodily ideal. Historian Matthew Hersch shows how the coinvention of flight suit compression garments for men and body-reshaping underwear for women cofashioned modern embodiments of technologically gendered power and constraint.[12] Cultural theorist Debra Benita Shaw argues that the classic spacesuit is not just an iconic representation of the autonomous, heroic, ideal Cartesian body but, in the context of space-age failures, also stands as an "ironic" symbol of how actual male bodies cannot transcend their own endangered embodiment and social dependence.[13] Architect Nicholas de Monchaux offers a compelling analysis of the Apollo lunar suit, examining its material and organizational composition layer by layer, detailing NASA's final selection of the Playtex company's "gossamer" adaptable suit made "out of underwear" over "military" hard suits that were built to spacecraft engineering specifications rather than the requirements of bodies.[14] De Monchaux argues, like Shaw, that the spacesuit is a social lesson—in this case, about the historically contextual limits and biases of *in*human systems engineering. A richly illustrated book by Smithsonian Institution spacesuit curator Amanda Young provides a typological history of spacesuits that is also a chronicle of Apollo-era suit material degradation and obsolescence, which reads metaphorically in the context of the failure of Americans to return to the moon or go on to Mars.[15]

Both de Monchaux and Shaw argue that the Apollo-era spacesuit's hard boundary-making design signaled NASA's failure to manifest a true spacefaring cyborg; however, ongoing spacesuit work within and outside NASA shows how a seamlessly technically integrated environmental subject continues to be a desirable institutional figure. Drawing on the feminist utopian view of cyborgic life, Shaw concludes that the Apollo-era spacesuit symbolizes an active refusal of the relational unboundedness and interdependence that cyborgic posthumanism offers.[16] De Monchaux notes that NASA's inability to achieve the cyborg developers' intentions for integrating any being into any milieu is a problem that continues to "haunt the systems architecture of the space age."[17] But as I argued in chapter 2, what is perhaps more haunting is the impressive lack, among NASA astronauts and scientists I spoke with, of expressions of desire for a fully comingled organic–machine cyborg body. I documented an occasional interest in biotechnology solutions for environmental adaptation, but these few mentions focused on shoring up barrier functions within existing purely organismal systems, such as making skin or DNA radiation-proof. I heard mention of clinical trials for drugs intended to mitigate the effects of radiation exposure, but no talk of biotechnical modifications. What I witnessed instead were redoubled efforts to create more sophisticated separation technologies. Therefore, while cultural critiques of NASA spacesuits are vibrantly historically accurate, they also make a certain kind of NASA spacesuit, astronaut, and systems sensibility seem fixed in time. The idea of "suiting" humans to space, which is the language of cyborg dreams as well as everyday NASA systems work, still takes the form of a better removable suit.

My appointment to meet a young Latina spacesuit engineer in iconic Building 7 took me away from the spacesuit museum foyer and halls; it also brought me into contact with separations of outside-NASA and inside-NASA attitudes toward spacesuited life as a changing form of life. Coming into Sara's office, I noticed two of the new cultural histories of spacesuits on her bookshelf, so I asked her if she had read them. "They look kind of interesting, but not really technically relevant. Plus

I really don't have the time to read them," she said, handing me a document she was writing on a new spacesuit system called simply J—the final name of this new Mars surface suit system had not been chosen yet. Sara spoke to me about problems with integrating new biomechanical and ergonomic requirements into the prototype Martian suit's hard, molded torso piece. Sara and other interlocutors spoke of the need for spacesuits to fit well and be flexible, demonstrating with their own bodies: gesturing, bending, shifting, turning, moving in empathy with astronaut bodies. And unlike in the past, female bodies had become standard kinds of engineering and astronaut bodies. Anticipation of changing spaceflight technologies and social inclusion practices was everywhere when I was at JSC. ("Exciting things are happening!" a white female radiation shielding specialist told me.) Astronauts spoke enthusiastically about participating in the design of new suits—"smart ones!" one young male as-can declared. Sara and her colleagues across JSC were working on helmets with virtual reality heads-up displays, computer mouse controls mounted on thighs, and fabrications with more "sensitive" materials. They were also employing new thinking about the physics of suit pressurization. While spacesuit technologies were expected to become "smarter" and more refined and flexible space/body separations, human factors and spaceflight training experts were also advancing ways to simulate more directly embodied extraterrestrial environmental experiences: manufactured space smells would acclimate noses to other ecologies, and "walkback tests" in which astronauts simulate long-distance planetary hikes in suits, as I describe shortly, would make suits more like athletic wear. In a large in-house meeting on Constellation's new human factors requirements, a senior female astronaut manager mentioned the need for "new design philosophies" to "devolve" the old Apollo-era planetary suits back to basics and then "rethink everything."

Some of this rethinking was happening as the Constellation-era quest for better systems integration revealed partial separation technology work focused on connecting the space between human skin and the spacesuit as a second skin. Most people I spoke with who were involved in spacesuit testing were engineers and human factors specialists responsible for, as one young white male engineer explained, the "skin out" side of the "skin in/skin out" separation of life sciences and human factors. While we sat in his cubicle one day, he showed me systems integration procedures he was revising that made him realize the organizational systemic rub happening at the skin level:

> We're trying to do a systems integration of various vehicle systems, but not even the human is integrated. One guy from our division sat me down one day. He was upset with me trying to integrate everyone. He said, "We deal with what happens from the skin outwards, we need to leave the medical doctors from the skin inwards." Well, I'm an intelligent guy, I've done work on the inside of the human, and I'm an engineer, and I don't think we need to separate that. The docs need to know human factors . . . and we need to know medical. Skin in/skin out is just based on your degrees. But we do need people to only worry about certain things, because not everyone can know everything.

The image of a systemic human whose interface with the environment happens at the level of skin emerges in both the theory of spacesuit design and the social organization of spacesuit work. In a technical evolutionary schema, the spacesuit is a barrier meant to—analogously—function like a second skin, where skin is understood to have evolved as an earthbound environmental separation system.

As skin-analogous extrabodily systems, spacesuits are not unique contemporary objects. They fit within a growing group of additive body-like parts understood to be parts of a functionally enhanced "better than well" bodily system.[18] These parts are autonomous because they can be connected to and separated from the body, like prostheses, without necessarily harming the body. But today these parts are also technically systemic in that they have body-autonomous sources of life-supporting materials or power, digital connectivity, and independent sensory or feedback-processing functions. Such extrabodily systems include military and environmental protective clothing as well as fitness monitors, smart watches, and virtual reality equipment. As extrabodily systems that are put into relationship with the body through the skin, spacesuits reveal how bodies can be technically theorized as partially connected to and partially separate from environments at the same time. Spacesuits are extrabodily systems meant to separate human bodies from harmful environmental conditions, but they do so by making partial separations between the interior body/suit and the exterior suit/environment. During Constellation, work on partial separations at the skin level took two forms: modifying existing suits and designing completely new ones. Both of these approaches reflected desires to incarnate astronauts as futuristic environmentally integrated beings.

## *Skin Out to Suit Edge: Spacewalking on Mars*

One day, in the huge JSC spaceflight systems simulation facility known as Building 9, I observed work on the partial separations of interior body/suit relations in a spacesuit walkback test. If ethical concerns circulating within NASA about the "image" of the astronaut as a biomedically at-risk subject cause bio*medical* testing to be hidden from public view, the visibility of the spacesuited astronaut as bio*mechanical* subject is a slightly different thing. When I worked as a research intern, I heard arguments between astronaut administrators and life scientists over the public release of photographs showing astronauts being subjected to medical experiments, a type of conflict that illuminates the status of the astronaut as a person who is simultaneously extraordinary and vulnerable. While astronaut bodies are highly protected, they also convey through gestures, comportment, and presence what Dominic Boyer has called the phenomenological "corporeality of expertise."[19] This produces the heroic but potentially tragic persona of the spacesuited astronaut, created through the tension between a superhuman ideal and the astronaut's actual experience. Both male and female astronauts whom I observed and in some cases came to know are, quite unsurprisingly, superconfident, capable, and sometimes arrogant, and they grapple with desires to stay safe and yet become intimate with outer space. Despite powerful guards against public biomedical testing, the spacesuited astronaut, as a technocorporeal unit, makes partly visible and partly invisible a schema of strategically partial body/suit separations.

Building 9 was a stop on the Space Center Houston visitor center's guided tram tour of JSC, and tourists passing through it on a glassed-in gangplank in June 2008 got to see the "walkback test" as a live event. The test is an analog that gauges the combined environmental "performance" of human and spacesuit as if the astronaut has been put in the risky situation of having to walk back ten kilometers to a vehicle or habitat on the moon or Mars. Through NASA's public relations strategies, every tourist is reminded of what space does to the unprotected human body. As visitors to Space Center Houston end their tour before JSC's memorial circle of trees planted for astronauts lost in the line of duty, a prerecorded message explains the sacrifices made in the name of the future. However, images of the EMU-suited astronaut in the visitor center act to immortalize settler colonialist dreams about people implanted in promising spaces, American enthusiasms for technology and nature, and triumphant claims staked in the name of territorial and environmental mastery. The live test that day in Building 9 was for a particular "suit and life-support system" being worked on for the Constellation project and its planetary operations and mission system integration scenarios. Building 9 also houses the International Space Station and space shuttle mock-ups used for training and equipment testing. The walkback test was one such simulation. I accompanied biomechanical engineers from the Habitability and Human Factors Division's anthropometry lab to watch this test of the Mark III spacesuit, an updated Apollo-era suit. I was unsurprised to find that one of the most veteran astronauts slated to do the test was a NEEMO aquanaut I had worked with on a mission.

The whole scene opened up a window into the deep ontological differences between suited bodies and unsuited bodies. As tourists watched from above, the male astronaut was rather unceremoniously zipped, snapped, and laced into a cooling-system long-johns-looking undergarment and then hoisted into the bulky, heavy Mark III spacesuit suspended from a framed treadmill-like setup called the POGO (partial gravity simulator). This pneumatic gimbal support structure was rigged—very imperfectly, my technical escorts assured me—to mimic the differential gravitational effects of different planets: one-sixth of the Earth's gravity on the moon or one-third on Mars. The suit and the human inside it hung within a tangle of sensor and monitor lines and equipment. Awkwardly, but forcefully, the suited astronaut began to walk rapidly, bouncing a bit in the rig, making the suit move. I had been told by an astronaut earlier that some in the corps were nervous about participating in these "suit validation tests" because of the risk of finding out something that would disqualify them for flight—another way in which

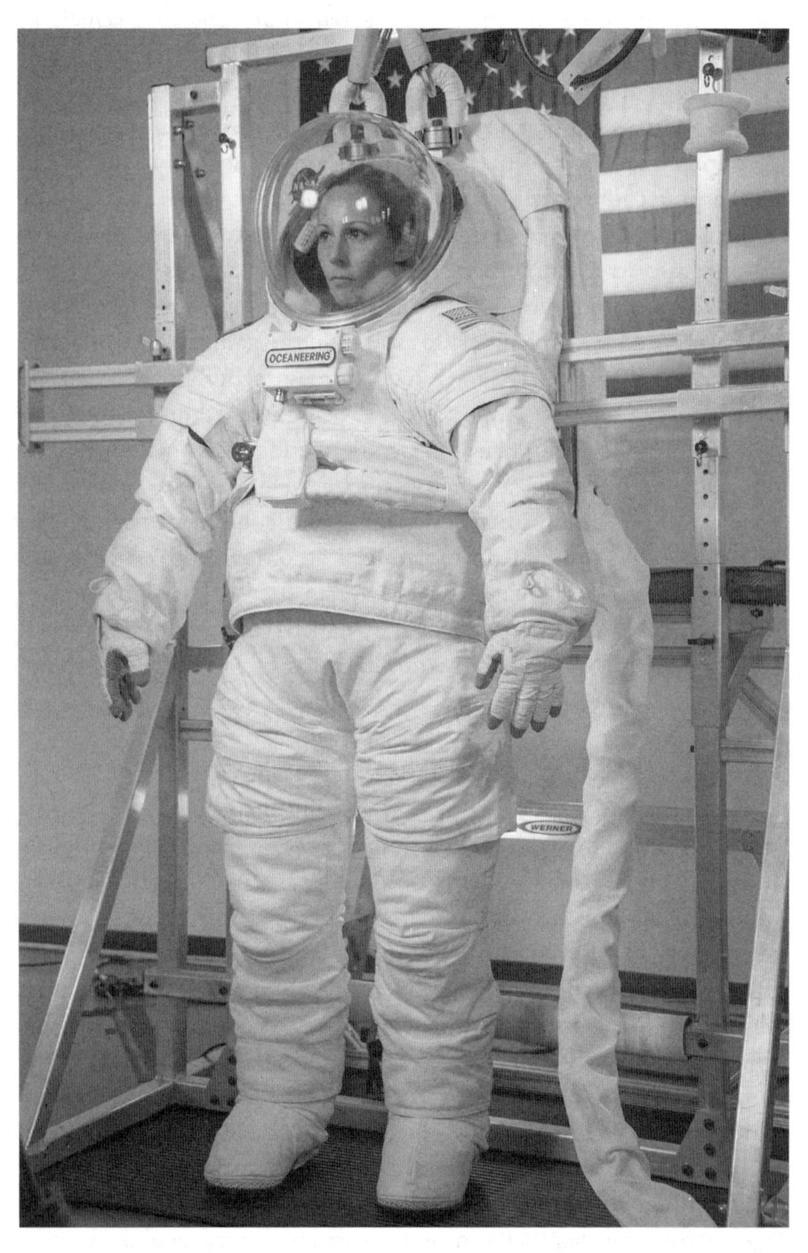


The NASA PXS "demonstrator" suit for testing fit and performance of the spacesuit as a partially separate extrabodily system. Courtesy of NASA.

spacesuits create status separations. In answer to my question about which astronauts were slated to be tested, a technician explained that the suit and test rig came in one size only, and as a result, only men were doing walkback tests at that time. As historian Amy Foster describes, the number of female astronauts able to spacewalk has been limited by the fact that, given the expense of redesign, fully pressurized extravehicular activity suits have been only slowly and incompletely modified for the sizes and shapes of female forms.[20] That has created another interior/exterior separation of astronauts, along gender lines, and an unequal division in social relationships with outer space.

As the walkback test generated data on body–machine–environment interactions, this information was organized to be disciplinarily shared or separated along the skin in/skin out division. On the treadmill, the astronaut smiled broadly through the fishbowl-like front of his helmet, looking like he was working out at a fitness center, big space boots stepping awkwardly. His trunk was covered by a hard, body-shaped core and fitted with a life-support backpack, his arms and legs made fairly flexible by accordion pleating. Video cameras recorded from different angles, and data from the astronaut/suit unit began to pour in on the computers of technicians seated and standing around the test site. Even in the aggregate, these data were understood to belong to either "skin out" body technology human factors or "skin in" life sciences categories: metabolic rate, ground reaction force vectors, carbon dioxide output, motion analysis measures, skin temperature, core temperature, heart rate, electrocardiogram patterns, periodic voice reports of skin-out discomfort or skin-in exertion levels. In addition, the whole suspension rig was calculating data on the astronaut's movements to compensate for falling, in an attempt to determine how a more effective and safe "center of gravity" could be engineered for the whole human/machine systemic unit. The toleration of uncomfortable hardware encasement and personally intrusive monitoring has a long history in NASA, mostly a gendered history of male suffering and adaptation to varying levels of bodily degradation and endangerment. The subjectivity of suffering created another ontological separation: I was told that the astronaut could choose to communicate privately with his attendant flight surgeon or openly with the assembled biomechanical engineers from the "extravehicular physiology" and "spacesuit systems" working groups. The test was completed, data were collected, and the suit was doffed to be donned by another astronaut on another day.

Despite the ontological divides that pressurized suits create, the idealized image of an extrabodily spacesuit system that can be worn, removed, and reworn by anybody builds the cosmological allure of this partial separation technology. The anonymous spacesuited NASA image of "every man" as a singular microcosmos is visible in the use of a spacesuited "Vitruvian man" on the EMU system patch. NASA historian Roger Launius describes the EVA suit's iconic power in these terms:

> Often described as a "spacecraft for one," space suits exist as highly complex, technical systems. [They] suggest our connections to our larger environment of Earth, the Solar System, and the universe. . . . [They are a symbol] that embodies dreams and beliefs about who and what we are, and what we may become. . . . These concepts are not just projected onto the material space suit, but are contained in its physical construction and invested in the astronauts who wear them. . . . Once in operation the physical object projects these philosophies onto the world around it; literally the space suit is a highly charged, metaphysical object that affects both the wearer and the observer.[21]

Launius warns that the Apollo astronaut in particular (and by extension the Apollo-suited astronaut) remains a "trope" that can also represent a "stalled" national vision.[22] But spaceflight advocates use suits to unstall perceptions of national failure and stagnation as well as social difference. I experienced this when I was given a suit to try on at a local spaceflight start-up company in Houston. "Try it on," an investor told the group of us, "this is about making space available to people like you all!"

While I was at NASA, this "one suit, many people" imagery fostered situated government-funded demonstrations of utopian nationalism. For a painting commissioned to represent the open future of U.S. Martian exploration, prominent aerospace artist Pat Rawlings asked Cheryl McNair, the widow of astronaut Ronald McNair, who was killed in the *Challenger* shuttle disaster, to be his model. Rawlings told me his intention for the image, in which Cheryl McNair, who is black, is depicted wearing a heritage-style spacesuit and kneeling down beside the dusty little Sojourner rover on Mars:

> And it is a thematic human piece; it's not really a technical illustration. It shows this explorer kneeled down next to the [Mars] Sojourner rover, wiping off the dust. And she's kneeled down there, and she's a black female astronaut and her badge says "Truth." And she's supposed to be the great-great-great-granddaughter of Sojourner Truth.

Rawlings's collaboration with McNair represents one of the many socially and ethically intentional projects that people have initiated inside an agency they consider to be a community or family, and that cannot be explained away as forms of governmental rhetorical manipulation. Nevertheless, the astronaut's representational identity as a future "Truth" renders the NASA pressure suit into a palliative container for a technical utopian desire to overcome a national history of enforced human bondage and traumatic separations. This complex set of intersecting desires and representational politics calls attention to NASA's role as an institution that produces visualizations of many kinds of systemic integration, often centered on technologies such as suit forms that enact social and political as well as technical tasks.

For extramural NASA researchers active during Constellation, such heritage envelope-style suit designs did not fully represent the potentialities of spacesuited life. Envelope suit designs are imagined as technically conservative and as failing to smooth out the problematic rub that occurs in the space between explorational bodies and technologies. These "personal spacecraft" suits mediate but also impede interaction between human and environment. In this way, they fail a core visionary goal of space exploration and biomedicine: to produce direct human–environment interactions that will constitute a transformative human milieu.

## *The BioSuit and Skintight Outer Space*

When I described the walkback test to some space life scientists and habitability designers involved in "human-centered" design weeks later, several of them suggested that I look at the NASA Institute for Advanced Concepts–funded speculative "biosuit" study.[23] My interlocutors considered this futuristic spacesuit vision far from usable but utterly fascinating and inspiring. In its theorization, design, and ongoing social life as a fabricated concept, the biosuit exemplifies a shift from the biopolitics of managing protected bodies to the ecobiopolitical production of manageable body–environment interfaces. The biosuit is also an attempt to enhance the social and environmental integration of spaceflight bodies in order to make space not only survivable but also recognizably and sensuously livable.

I first encountered the Bio-Suit System project as a PDF-format "paper NASA" report written by an MIT team of professors, students, and contractor collaborators; today, the more coolly named BioSuit is a media-attracting prototype artfully worn by inventor Dava Newman, who became the deputy administrator of NASA during the Obama administration. Newman's MIT project, which was developed in collaboration with space architect Guillermo Trotti and the Dianese motorcycle leathers company, draws inspiration from the midcentury formfitting

A rendering of the NASA BioSuit. Other images are associated with the proprietary entities involved in the development and promotion of such a suit. Courtesy of NASA.

flight suit rather than from the gas-filled heritage spacesuit technology. Like documents on early Apollo designs, the Bio-Suit System white paper reveals the logics of technical desire. Its promotional images situate U.S. spacesuits within a speculatively creative and collaborative space, less driven by national competition and boundary keeping and more open to the worlds of science fiction, artistic fabrication, and architecture. The Bio-Suit System creates a more partial separation from outer space and shifts the edge of that separation. It moves that edge from outside the spacesuit (heritage suit) to outside a second skin (BioSuit), creating an extrabodily system that is not skin covering but skin enhancing. According to its advocates, the Bio-Suit System relates individuals and social groups within space environments better than does the pressurized "spacecraft for one," reinforcing the engineering ideal of systems as evolving technologies. It also manifests the desire of designers to make an extrabodily extension, not a bodily container.

The designers of the Bio-Suit System claim, in a 2001 report, that the suit has an interdisciplinary advantage over the Mark III spacesuit's strong separations between medicine, biology, and engineering:

> A *Bio-Suit System* stands to revolutionize human space exploration by providing enhanced astronaut extravehicular activity (EVA) locomotion and life support based on the concept of providing a "*second skin*" capability for astronaut performance. The novel design concept is realized through symbiotic relationships in the areas of wearable technologies; information systems and evolutionary space systems design; and biomedical breakthroughs in skin replacement and materials. By working at the intersection of engineering; design; medicine; and operations, new emergent capabilities could be achieved.[24]

In the "Cyborgology" section of the report, in which they locate their design as a revision of an abandoned NASA intention, the authors remind readers that the cyborg concept suggested the development of not only internal (medical) adaptations but also "exogenous" ones that could act like the enhanced skin the BioSuit aspires to be. The Bio-Suit System designers expand the scope of the "environment" that teams must address, making it a problem space for what they identify as "anthropological" problems:

> If space exploration is going to *transgress* existing possibilities via the Bio-Suit effort, anthropological issues must be investigated and addressed through the design of such a system. We must re-evaluate the role of the astronaut as an explorer and as a representative of our species on earth. We must carefully design the physical, psychological, and cultural environment that future explorers inhabit during these explorations and truly understand what we are trying to gain from the experience.[25]

Although it seems likely the authors mean "transcend" here rather than "transgress," both words evoke space advocacy's goal of refusing given assumptions about risk and the threatening nature of extremes, a goal the team calls "flexibility for the future." This discourse about technical advancement reflects contemporary rhetoric surrounding the usefulness of all kinds of extrabodily systems.

The BioSuit does not just contain the body, it adds technical elements to the body's integumentary system and turns its muscle system inside out. Its bioengineered elastic "skin" and "exoskeleton" of tensile strings interface with the extraterrestrial environment, restoring the physical, social, and "anthropological" bodily integrity lost within a pressure suit. Instead of putting the body inside a pressurized volume that creates resistance, torques against movement, and damages the astronaut's skin, the BioSuit is a "second-skin system." This system starts with a spray-on protective film that the astronaut applies to naked skin before donning the skintight suit, which selectively controls the passing of molecules between environment and skin, allowing the BioSuit to sweat, cool off, and heat up as needed. The suit resurrects the midcentury jet pilot's tight heritage flight suit, but this is not simply the body shaper form that historians link to gendered power divisions. The BioSuit applies protective "mechanical counter-pressure" to the astronaut's body, against but also in relation to the vacuum of space. Embedded in the suit are audio and tactile "information" technologies that enhance the astronaut's direct environmental awareness. The wearer's skin and other bodily systems, such as joints, are "augmented" with prosthetics and "biomimetic locomotion algorithms" that enhance strength and locomotion.[26]

The *bio* in the BioSuit design schema is not a quality but a set of functions that relate bodies to environments through evermore selectively partial separations. The design moves "bio" functions from inside the organism to its outside, at the site where organism and environment meet, remaking the life–environment boundary. Gone are the routine references to the medical problems and design challenges that the pressure suit's "microenvironment" presents—that term does not appear at all in the Bio-Suit System study.[27] Instead, the study focuses on the problem of microspatial skin–environment interfaces and "exchanges," where "skin" is operationalized as either "grown" biology or "built" technology with complementary "performance" capabilities. Unlike pressurized suits, the BioSuit moves to the next ecobiopolitical level, actually exposing small amounts of human skin (areas of 1 square millimeter) to the space environment. The designers claim that they have created a body–material surface interface that qualifies the BioSuit as a "wearable computer."[28] The BioSuit's sensors enhance the capacity of human skin to communicate information about interaction and creates a digital memory of interaction.[29] The BioSuit is an ecobiopolitical strategy to mitigate risk, enhance human–environment interactions, and make those interactions virtually transparent to the one who inhabits the alien milieu with a desire to belong there. Mechanical "life support" is exteriorized to portable breathing technologies. Although its designers call it the *Bio*Suit, the suit is more like the "environmental suits" found on *Star Trek* and *Star Wars* fan sites, which replace the old term *spacesuit.*

In order to intervene in an ongoing argument about whether humans or robots are better suited for the exploration of space, the designers emphasize that their second-skin system blurs separations between human and robotic capabilities. They write, "The explorer is hardly aware of the boundary between innate human performance and robotic activities."[30] By making it safe for humans to perform things in space that only robots could do before, they view their work as humanizing, individuating, and socially enhancing processes of outer space environmental integration. Referencing an ongoing objection astronauts have to being "encased meat" in their spacesuits, the designers argue that the BioSuit "change[s] the exploration paradigm from 'spam in a can' to an individual interacting with and inhabiting an extra-terrestrial environment."[31] The report includes a "Visualizations" appendix, which contains an illustration of two BioSuited humans in a vital social environmental interaction—a handshake:

> In order to promote the type of research required of a Martian colony, personal interactions must be facilitated. Here we see two researchers finalizing an agreement in much the same way we would on Earth. Due to the low profile of the Bio-Suit and the increased mobility and tactile sense, interactions such as a handshake become possible and pleasurable. This demonstrates how the physical design of a space suit can directly impact the daily experiences of the wearer on a utilitarian, psychological, and emotional level, thereby promoting overall health and happiness.[32]

The BioSuit wearer, then, is not just technically but also socially functional.

In this schema the idea of an evolving human bodily system made up of biological organs is interrupted in favor of an assemblage of organic and inorganic systems that are understood to be both structural and vitally emergent. The capacity of the BioSuit to advance new forms of biotechnical systematicity calls to mind how Deleuze and Guattari chose the term *system* as the neutral descriptor for their conceptualization of "the machine" as any constructive set of heterogeneous relations. For Deleuze and Guattari, "the machine" can be a "system of interruptions and breaks" that ensures continuous dynamic change or a "system of means in terms of the aims of desire."[33] While these theorists use the term *system* liberally in their work, they are also known for rejecting it. What they reject is the old concept of a closed system in favor of a porous, morphing one. The BioSuit, with its membranous skin enhancement, is also a machinic system designed not just to protect but to open up to space and desire.

The BioSuit remakes the spacefaring body into an aesthetic and desirable form. Its skintightness performs intentionally transgressive technoerotic power. In this way, the BioSuit separates itself from NASA's history of keeping spaceflight sex and desire hidden. Unlike the awkwardly embodied Mark III walkback test I witnessed, getting into and wearing the BioSuit looks inviting and canny. Artists' renditions of the suit and photos of Newman modeling the suit over the course of more than a decade reproduce similar effects: female or indeterminately gendered bodies are posed in ways that mark a break from the balloon-like vaguely humanoid shapes of the old suits. The BioSuit's fitness-wear look emphasizes its capacity to remold human bodies while fulfilling its purpose as space expeditionary equipment. Photos of Newman posing in this inside-out suit before rocket engines that look like artist H. R. Giger's veiny living machines, or seated as a kind of cosmic bodhisattva, connect her suited body to transcendental erotic genealogies, including New Age iconography and sci-fi and fetishistic fashion. The suit also decouples exploration from its Apollo-era association with heroic sacrifice and suffering, reorienting it to the possibilities of extreme enjoyment afforded by the pleasures of flexible design.[34]

Making NASA's reference to Leonardo da Vinci's Vitruvian man seem antiquated and abstract, the BioSuit is a piece of wearable bioart in which biomedical risk appears duly retired and alien environments are comfortably and vitally inhabited. In an arty photo of Newman in the BioSuit taken on the MIT campus, she is perched, legs crossed, on a futuristic sculpture, Henry Moore's *Three Piece Reclining Figure: Draped* (1975). This work, an MIT sculpture guide explains, is designed to "assimilate" into its surroundings and to signify a reclining nature goddess, representing tensions between the built and natural environments, fullness and emptiness. Fit between mass and void, occupying both negative and positive space, the BioSuited female body is functionally separate but poised to integrate, interact, and engage, in multiple senses of those terms. Flexible but tightly wrapped, the astronautical body is exposed through a stylized biophysics that takes the human out of what the original cyborg designers called the pressure-suit "fishbowl" and into a less mediated, but in fact more thoroughly controlled, interaction with natural and artificial milieus. Embodied confidence emanates not just from the astronaut herself but from the whole suit design, making it illustrative of an "integrated systems" future in space environments that are more thoroughly social, nascently reproductive, recognizably human, and potentially habitable than they were last century.

The BioSuit thins the border of inside/outside separations, making them really a matter of insides only ever partially separated from other insides. Outer space represents the modernist utopian "outside" and "beyond," but when humans prepare to live there, space forces confrontations with ideas about discrete outsides. In trying to manifest the *res extensa* promised by science and technology in "our action travel around the blue planet," as Bruno Latour described orbital spaceflight in an address presented at a design college, the actual difficulty of space travel makes it appear to be a "shrinking enterprise" haunted by a growing concern that there is "no outside left."[35] However, the BioSuit is an ecobiopolitical artifact of ongoing attempts to deny, accept, manage, and exploit that concern all at once, rendering it into an "enticing interface." This is a phrase David Howes uses to describe a goal of midcentury designers like aerospace contractor Raymond Loewy who set the "sensual logics" of late industrialism.[36]

As elite separation technologies that foster a futuristic extimacy with forms of environmental protection, the superseparate EVA pressure suit and the sensual BioSuit are extrabodily systems that stand in stark political contrast with the partially protective spacesuit-like clothing worn by industrial laborers. Dvera Saxon addresses this contrast in her analysis of industrial agriculture's ecobiopolitical control of strawberry fields and fieldworkers through the production of inadequate partial separations, such as spacesuit-like garments that meet basic requirements for daily protection but do not mitigate extreme lifetime exposures.[37] In light of this, the life-preserving spacesuit stands in relation to aerospace's contributions to processes that create large-scale industrial contamination in service to national productive intentions. In the next sections, I situate spacesuits within the larger context of NASA separation technologies that attempt to reconcile technical as well as ethical problems with spaceflight's industrial systems.

## Clean Machines: Making Interplanetary Systems

At JSC one winter day, I stood with a veteran lunar scientist watching people in laboratory cleansuits work with unearthly rocks. Housed in a nondescript building, JSC's Lunar Curation Laboratory is a space of segmented see-through separations. In this space, storage containers and gloveboxes allow for sterile manipulation of outer space matter. The facility is managed by NASA's Astromaterials Curation Office, which is responsible for the safeguarding and distribution of moon rocks, stardust, meteorites, and, someday, Mars samples. It was a quiet day there. Garbed in white overalls with hair-covering caps, two male lab techs were busy with machinery, but not with moon rocks. There were a couple of rocks on display for lab visitors, chunks of gray rock infused with an otherworldly mystique sitting in their sterile containers. Leo, an older white male planetary scientist with decades at JSC, relayed the history of the lab to me in his inimitable way, blending his voluminous store of selenological facts with personal observations about the meaning of Earth–lunar relations. Leo's colleague Wendell Mendell—NASA's chief lunar scientist during Constellation—described the relations between Earth and the moon to me as so close the two "almost" form a "double-planet system." Mendell wrote a paper for *Space Policy* declaring that the moon and Earth "combined" constitute a "Greater Earth system" that has implications for thinking about the "weak" boundaries of earthly space.[38] This is an argument echoed by Wendell's colleagues at the International Space University in Strasbourg, where a team of instructors and students named their lunar habitat design proposal the "Luna Gaia closed-loop system."[39]

With such relations in mind, spaceflight experts work to provisionally connect systemic interplanetary relations. This involves the production of technical systems that allow for control of those relations according to expert rules and desires. There are many partial separators involved, but here I survey one kind: the "clean" robotic spacecraft. Clean spacecraft are intended to separate biological systems *but* provisionally connect environments. Even as they enable such connections, clean spacecraft also represent a separation between the spatial management responsibilities of engineers and those of environmental ethicists.

Like the astromaterials lab cleansuit, the clean spacecraft is a partial-separation technology for making interplanetary systems; by preventing undesired *bio*logical exchanges, these craft enable controlled *eco*logical systematicity. As nonbiological but socially animated systems, clean spacecraft—real or imagined—are objects that materialize U.S. interplanetary relations. To date, however, scholars interested in robotic spaceflight's relations with planets have focused on technologies rather than on ecologies. Janet Vertesi describes how Mars robotic spacecraft operate as extensions of scientific sensoria that enable forms of interplanetary knowledge through "technomorphic" associations of bodies and machines. Mars mission social scientists, project scientists, and managers write about how Mars mission planning is an act of interplanetary social organization.[40] But part of the social organization of interplanetary relations involves "planetary protection" separation techniques. In the Lunar Curation Lab, the cleansuited job of making uncontaminated cosmic material samples and distributing them to global recipients is an example. Such protection techniques also include manufacturing tools and practices that ensure that spacecraft headed for planets targeted for astrobiological life searches are as free from "forward contamination" by earthly microbes as possible, as well as procedures for protecting Earth from "backward contamination" by extraterrestrial sample return.

During Constellation, the problem of what planetary protection means in a new human spaceflight program also shed light on ongoing separations of spaceflight engineering and environmental ethics. While I was doing archival work in the NASA Headquarters history office, I ran into Michael Meltzer in the stacks; he was working on completing his history of NASA planetary protection practices, *When Biospheres Collide*.[41] He argues that what is being "protected" is not planetary life but life science itself. Protocols that maintain biological boundaries through technologies like sterile spacecraft act to certify scientific claims about the presence or absence of extraterrestrial life. This supports Stefan Helmreich's argument that what is partly at stake in astrobiological searches is biology's potential to become a cosmic-scaled science like physics or chemistry, whether through the discovery of different forms of life beyond Earth or the discovery of similar forms.[42] However, even as clean spacecraft reinforce the dual universal authority of engineering and science, they also separate the meanings of "protection" as a physical or ethical act. On multiple occasions, I was vehemently informed by U.S. and European planetary protection experts that "planetary protection isn't environmental ethics!" My attempt to interview spacecraft cleaning engineers about their thoughts on whether or not earthly environmental protection ethics should be extended to other planets met with confusion, with one technician telling me that his job was "just" to "meet the engineering requirements."

Even as clean robotic spacecraft are made to act as proxies or precursors for human exploration by creating controlled partial Earth–space relations, they also make legitimate separations between technical systems and moral systems. During Constellation, I watched as scientists worked on astrobiological, colonization, and terraforming system designs, but ethical questions were kept in abeyance. Gatherings focusing on planetary protection and spaceflight environmental ethics still often occur separately from or separated within larger space science and policy gatherings.[43] In their Constellation-era article "Separated at Birth,

Signs of Rapprochement," which appeared in the journal *Ethics and the Environment,* philosophers Erin Moore Daly and Robert Frodeman take the Deweyan perspective that spaceflight environmental science and ethics will stay separated unless philosophers learn to engage in "pragmatic" policy problems, such as whether or not to enforce the protection of lifeless spaces.[44] This highlights an ongoing separation in the technical and ethical dimensions of systems thinking and practice.

As designs for what would be astrobiologically *un*clean but biologically lively human-inhabited spacecraft were being aimed at Mars during the Constellation program, I watched astrobiologists and human spaceflight advocates articulate different valuations of Mars as an ecological system. At the first Constellation-era NASA lunar science conference at the Ames Research Center in Mountain View, California, in 2007, a veteran astrobiologist told me that "the moment human boots hit the ground" on a planet it becomes astrobiologically "unusable." I watched people continue to debate the ongoing problem of extraterrestrial biological futures at the 2016 "First Contact" science and science fiction conference on alien encounters. These debates pivot on whether humans qua cosmic life have an ethical obligation to enliven other planets. Humans could do so by starting human colonies or sending spacecraft to foster indigenous or exogenous microbial ecopoeisis, a process developed by Penny Boston and her colleagues. Boston, a cave microbiologist and currently the director of NASA's astrobiology program, described to me her lifetime commitment to advocating for the microbial life that social worlds ignore but vitally depend upon. Within spaceflight circles, debates continue about whether or not "dead" or "contaminated" planets and other bodies, like Earth's moon, have the intrinsic right to remain undisturbed.

But within these debates, there is never a question about whether those bodies exist outside a larger "system." As Daly and Frodeman conclude: "Abiotic nature can also have value through the *relatedness* of nature and natural objects to human beings."[45] Whether the ethical reasoning is bio-, eco-, or geocentric, the intrinsic systematicity of the solar system confers a moral imperative on human systems maintenance, building, and boundary work—whether just on Earth, as ecosystem historian Frank Golley argues in his anti–space colonization essay in a definitive book on solar system environmental ethics—or through spaceflight.[46]

Also at stake in these arguments about the future of spaceflight environmental ethics is the increasing load of spaceflight waste on orbit and on planets.[47] All spacecraft have come to represent the classic anthropological definition of "pollution"—matter out of place. In transmitted photographs of the moon and Mars, once-active vehicles have become nonfunctioning systems, stopped, stuck, and abandoned in situ.

National imperatives to systematize outer space through processes of controlled connection and separation gave U.S. spaceflight institutions carte blanche to leave industrial waste across the solar system, but contemporary political pressures for systemic accountability have led NASA to "clean up" its earthly environments. Although I never encountered anyone within NASA who was familiar with systems engineering or management ethics, one version can be found in the work of Quaker-schooled systems management advocate C. West Churchman, who served as space sciences director at the University of California, Berkeley, in the 1960s. His once-popular book *The Systems Approach,* first published in 1968, warns administrators and scientists about the consequences of not folding civic accountability into systems management. Churchman reminds "rational" systems thinkers not to consider themselves as "apart" from the systems they work on and advises them to identify "whose objectives" the managed system must serve as well as "the customers of 'the system,'" which, for government agencies, is "a subset of the citizens."[48] In his subtle but selective partial separation of the body politic into systems customers and others, Churchman reveals the embedded elite ethical partiality of managed systematicity.

While I was in the field, NASA administrators were reformulating their environmental management strategies and addressing histories of contamination at sites like JSC's White Sands (New Mexico) Test Facility, which has been used for testing "system behaviors in hazardous environments," and at the Jet Propulsion Laboratory near Pasadena, California.[49] These were not simply environmental remediation activities but were also part of NASA's efforts to control its sometimes fraught civic relations as a producer of planetary pollution *and* planetary systems knowledge. Contemporary efforts to clean up water contamination caused by JPL activities in the Pasadena area involve both discursive and material separation technologies that make systematicity work.

## Civic Filters: Toxic Water and Selective Partial Separations

In U.S. spaceflight's political geography, Southern California is the technical and economic capital of robotic spaceflight. Most of the clean interplanetary spacecraft that maintain planetary biological boundaries and consolidate a technical ecological solar system are managed at NASA's Jet Propulsion Laboratory, situated in the forested hills of La Cañada Flintridge, near Pasadena. As at JSC, the site is full of wildlife and people sharing the campus's pathways and grounds. Now part of the California Institute of Technology, JPL grew out of entanglements between outlandish local aerospace companies and midcentury government and military operations. Its organizational history is built from the work of occultist and illicit-drug-using rocket engineers as well as the staid bureaucratic functionaries and legions of space-loving engineers and scientists who come and go.[50] With its website's pledge to "bring the universe to you," JPL has operated some of the NASA programs and crafts most beloved by the public, from the interstellar Voyagers, equipped with human cultural signs and sounds meant for what were assumed to be biologically and culturally analogous aliens, to the Mars rovers, which broadcast unforgettable images from the red world's uncannily familiar-looking surface.[51] As a spectacular work site, JPL is as much an element of Los Angeles's claim to imaginative otherworldly production as Hollywood. But while its rovers worked to find traces of life-sustaining water on Mars, JPL also became the site for earthly groundwater decontamination work. JPL's locally sited separation technologies filter contaminated water and create social responsibility distinctions between the center's current civil spaceflight and past military aerospace institutional embeddedness as a part of the Los Angeles area. Yet JPL scientists and engineers also work to more strongly connect the spaceflight exploration systems industry with earthly technical industries.

NASA's concerns with human health contaminants span Earth and Mars, but its systems work in these sites is kept partially scientifically separate because of the sites' differently constituted political ecologies. For decades, JPL has designated institutional teams to implement the terms of an agreement with the U.S. Environmental Protection Agency to remediate JPL's contamination of local water sources with volatile organic compounds, or VOCs. Chief among the contaminants is perchlorate, an element of solid rocket fuel. When the U.S. Army ran JPL from 1945 through 1958, it dumped chemicals from the testing of jet engines and rockets into dozens of "seepage pits." From the early 1990s, JPL has deployed more and less successful systems techniques to separate VOCs from local water supply systems and dispose of the contaminants elsewhere. NASA employees view this cleanup as an army responsibility borne by NASA, adding to the list of things and processes that interlocutors cited to me over the years as evidence of NASA's hybrid identity as a civilly oriented institution with only partial connections to the military. But the social and political partitions between kinds of systems knowledge and work extend inside NASA itself. Perchlorate is a well-known aerospace contaminant on Earth, but when the Mars Phoenix lander identified the chemical in Martian soil in 2008, the principal investigator for the Phoenix project described it as a "surprise" and as such an unfamiliar substance that "all of us had to go and look it up."[52] JPL employees associated with the local groundwater cleanup are, for the most part, not involved in spaceflight work, and the public documentation JPL produces on its "cleantech" activities emphasizes the agency's civic commitments but does not highlight a relationship between its earthly environmental filtration work and extraterrestrial environmental knowledge production. As spaceflight interlocutors, including legal experts, complained to me throughout my fieldwork, NASA attempts to control its legal issues and keep them apart from images of its spaceflight activity.

The environmental systems work separations at JPL fit into a larger context in which NASA keeps its Earth–space knowledge production and operational decisions partially connectible and separable. One of the domains in which this is visible is the separation between NASA's work on Earth's climate and its work concerning other planetary climates. NASA satellites monitor the impacts of anthropogenic climate change on Earth, but the agency also supports work to consider the social and ethical benefits of altering the ecologies of other planets. The discovery of perchlorate in Martian soil prompted a 2014 workshop at NASA Ames intended to foster terrestrial *and* spaceflight collaborations on the topic of the potential harm and usefulness of perchlorates in the human inhabitation of Mars, since perchlorates can be used as transportation fuel. On the workshop's agenda, leading astrobiologist Chris McKay was scheduled to speak on the concept of "ecosynthesis," which has become for many astrobiologists the successor to the concept of "terraforming," in part because it includes humans as ecological elements in planetary ecosystemic transformation. McKay argues that if there is no indigenous life on Mars, human-instigated ecosynthesis there would require the same processes by which industrialization is producing global climate change. For McKay, ecosynthesis that fosters indigenous life on other planets or transports terrestrial life there can constitute an ethical act of spreading "diversity" of life forms in the solar system.[53] In a paper included in an ethics module on the Earth System Science Education Alliance website, McKay observes that all environmental ethics "systems" are based on principles concerning the relationship of nature and life; Mars presents the problem of choosing one over the other, and "life" should win.[54] This perspective on astronautics as a technical form of ecological systems building was common among people I worked with and interviewed across my field sites, where varieties of environmental ethical sensibilities coexisted with commitments to expanding U.S. spaceflight industrial systems.

Even as agency leadership enforced NASA's political and legal interests in partially separating its Earth and space environmental work, astrobiologists like McKay and other life and biomedical science interlocutors shared a belief in spaceflight's civic duty to rethink and remake systemic life. One day at JSC I sat with Kyle, the flight surgeon I touched base with often, in a coffee shop on Saturn Boulevard, down the road from his clinic office. I told him that I was interested in spaceflight practitioners' responses to earthly environmental problems. I noted that space physicians deal with futuristic molecular-level environmental problems in real life and in science fiction, citing a BBC mockumentary about a long-duration space voyage in which an astronaut sick with radiation-induced cancer commits suicide to avoid contaminating his crewmates' closed-loop water system with the chemotherapeutic drugs in his urine.[55] As we sat amid the midday caffeine-refueling rush of JSC employees and contractors, Kyle agreed with my observation that terrestrial institutions are now joining NASA as customers for solutions to the interconnectedness of micro- and macroenvironmental "loops." What used to count as problems unique to spaceflight suddenly seem relevant on Earth. He went on:

> And in a way that's a kind of Copernican inversion. I mean in the old days you were limited in, you took your garbage and you threw it away, and now we know, there's no away into which one can throw it. Okay? . . . Copernicus was the first guy who took the same observations [that others did] and all of a sudden realized, wait a minute, this means the sun is the center of the universe. . . . It's the same data set, but the interpretation, that's called a Copernican inversion. And that's totally different than a paradigm shift. A paradigm shift is a minor course correction. A Copernican inversion is an absolute breakthrough.

In the months that followed this conversation, I observed space habitat designers faced with the problem of how to make the necessary separation of bodies, suits, spaceships, and planetary environments more functionally integrated. In one meeting, I was handed a sketch in which spacesuits were also airlock ports; in other words, the connected but separable suit-door components were both walls and skins, both protective

NASA sponsored a public participatory "spacesuit design challenge" vote in 2014 to select the "cover layer" of its "evolving" Z-2 Mars exploration suit. "Technology" layer (left) with "light-emitting patches" that could be used for crew member identification won over the "Biomimicry" design (right) symbolizing the evolutionary environmental adaption properties of ocean animals and the "Trends in Society" design (center) representing future "sportswear and the emerging world of wearable technologies." (For the Z-2 prototype challenge results, see NASA, "NASA's Next Prototype Spacesuit.") Courtesy of NASA.

clothing and doorways. Scott, a spacesuit designer, explained: "You step in, suit up, disconnect, and walk away from the habitat." In this Copernican inversion, suited human bodies unsuited to extraterrestrial environments are not boundaries but vehicles for selectively breaking down social understandings of insides and outside. Connecting and disconnecting are both systemically relational acts of, as spaceflight advocates would say, "transition to being a spacefaring species."

The political ecology of NASA's efforts to connect Earth and space environments is powered by changing technologies of partial separation. Changes in spacesuit forms and philosophies mark changes in spaceflight capacities to put humans into more desirable relationships with outer space, as part of a trend in the terrestrial production of extrabodily systems to mediate human–environment experiences. But spacesuits are also political-ecological indicator technologies. As spacesuit extrabodily systems go from full to partial body/environment separations, they evidence how environmental hazard protections can become more normalized in everyday spaces. And while planetary protection techniques keep Earth and space apart biologically, they also make it possible for environmental engineering practices and environmental ethical concerns to remain in separate social and political spheres. These social dividing practices allow NASA as an agency to control its political systematicity—to produce data on industrial climate impacts but also to be fueled, funded, and technologically enabled by national and industrial systems-building interests. In all, NASA workers are engaged in rethinking and remaking considerations of life's systematicity from one end of the human world to another. They build a variety of environmental, political, and cosmological perspectives, and travel within and outside the agency's boundaries, as members of communities of practice and inhabitation. In the next chapter I focus on the distributed communities engaging in space habitat systems design, in which the act of making space livable becomes infused with environmentally transitional social and political aspirations.

# 4

# TRANSHABITATION

## UNIVERSALIZING ARCHITECTURE

During the Constellation program, JSC's Building 29 housed ideas for how humans could transition from surviving in space to living there. It contained sketches of spacecraft interiors, models of habitats, and, deep inside the facility, a life-size inhabitable machine. Installed in a cavernous room that once held a huge centrifuge used in the study of g-force effects on human subjects, this habitat-machine was a complex of interconnected cylindrical modules lined with gleaming conduits. You could enter it through an open hatch. When you stood inside it, the Bioregenerative Planetary Life Support Systems Test Complex, or BIO-Plex for short, looked like a science fiction movie set.

Built as a space analog in the late 1990s, BIO-Plex is kin to other twentieth-century closed-loop system habitats. Archived reports on the project describe how the habitat was meant to be closed up with a crew inside, as if on transport to another planet. The human crew would tend living and mechanical "integrated systems" to create an environmentally renewing space. A system of growing plants and microbial colonies would live off the crew's waste and labor and in turn supply food and oxygen for a "human-in-the-loop" dwelling powered by electricity. Calling this system-of-systems machine a "habitat" conjoins astronautical and ecological ideas about the form and evolutionary purposes of living spaces. Teams of people empowered to do so design these systems to produce the relational conditions of inhabitation anywhere.

When I was at JSC, BIO-Plex was in suspended animation, its airlock doors permanently propped open to institutional space-time. But its interior spaces were alive with habitation design experiments being pursued to make Constellation's new crew spaceflight capsule, Orion, comfortably habitable as well as to enable planning for future lunar and Martian surface habitats. While BIO-Plex's meticulously engineered hoses and conduits stayed quietly empty of the life-support gases and fluids they were designed to carry, people came and went through its hatches, filling its spaces with in-process design projects. It was strewn inside and out with fabric swatches, bits of wood, and tools being used to build full-scale models for living spaces of different shapes and for the configuration of sleeping and dining arrangements. Like other aspirational semiclosed-loop ecological systems built in the twentieth century, BIO-Plex has yet to be occupied by test crews as planned. Yet some of its design elements can be seen in the International Space Station and in plans for other extreme environment habitats, such as terrestrial disaster

BIO-Plex interior. Photograph by author.

shelters and polar research shelters, and in architectural designs for soon-to-be built underwater hotels. Like other multimillion-dollar NASA one-off experimental systems, BIO-Plex was subject to policy shifts and funding cuts, and it ended up as a transitional object—a placeholder for spaces and ways of life to come. In design parlance, Building 29 and BIO-Plex became sites for "concept work," "mock-ups," and "prototypes." As scholars of science and technology have shown, modeling and prototyping are used to work out situated theorizations of sociotechnical relations.[1] But BIO-Plex was also a hyperelite government housing demonstration project. Operating within historical networks of federally and privately funded started-up and shut-down projects, people continue to work to make space habitation's moving and morphing parts. They are working out, as I show in this chapter, universalistic designs for inhabiting and controlling environmentally systemic spaces with a terrestrial home-to-cosmic-home scalarity.

## Transitional Practices

When I spent time with space architects and designers who were using workspaces inside and outside NASA as sites for generating ideas for space capsule and habitat design, the problem at hand was not just how to build habitable containers but also how to produce habitability out of systemically related organisms, things, and spaces, an aim that was engineered into BIO-Plex's name. One day, while sitting inside a BIO-Plex module mocked up as a moon base dining room, with table, chairs, and pretend windows displaying lunar landscape photos, I asked senior JSC architects to define habitability. Justin, one of two middle-aged men at the table with long histories in working on extreme environment habitation, slowly shifted in his chair. He replied wearily, "It's *soooo* messy." The messiness, he and others explained to me over years, had to do with working out the parameters of habitability in engineering, architectural, biomedical, industrial design, and cost terms during constant program changeovers in an agency struggling to define spaceflight's purposes. Justin and his colleague Phillip had been part of a team that worked a decade earlier on NASA's renowned TransHab, an inflatable "transit habitat" designed to attach to a spacecraft or sit on a planetary surface. After an initial infusion of funding, the project made it only to the prototyping phase. Despite calls for its continued funding from spaceflight advocacy organizations like the National Space Society, congressional battles over mounting International Space Station costs resulted in the White House's blocking funding for TransHab and related projects. When I was at JSC, a dollhouse-size TransHab model sat on the floor in the foyer of Building 29, left as a memorial to what a veteran black civil servant architect and engineer referred to, with a resigned tone, as the project's "life and death spiral," which haunted projects he was currently leading.

In addition to the term *habitat,* TransHab's moniker foregrounds the significance of the *trans-* prefix in spaceflight systems work and discourse. In this work, habitation is a process of transferring and transforming in spaces where inside/outside, body/habitat, habitat/environment distinctions and relations can be systemically transfigured. The valorization and normalization of the *trans-* prefix in this contemporary scientific and technical context stands apart from the ways in which *trans-* can configure experiences of exclusion and inclusion for communities of transitional sexual and gender identity. This prefixed distinction highlights the power of elite practices to selectively connect and separate the terms, and worlds, of social change.

Spaceflight transhabitation processes, passionately pursued within and around JSC while I was there, connect modernist and environmental technical perspectives on built habitats as systemic units of ecological transition, material transformation, and ideological translation. In the United States, space habitats are dwellings positioned at the intersections of technocratic and ecological discourses. But they are also dislodged from the given terrestrial conditions that implicitly undergird theories of dwelling. Intentionally more-than-earthly habitats, they display reasons, modes, and trajectories for existential change. Like modes of building verticality and modularity that operate in global settings today, transhabitation work in U.S. space programs materializes certain kinds of transcendental ecosystemic sensibilities. It does so by advancing assumptions that the conditions of existence, even extreme ones, are controllable and transmutable. Houston architect Constance Adams, who worked on BIO-Plex and TransHab, praises the value of space habitats designed to enable "reconfiguration as necessary over time."[2] Phillip, who worked with Adams, told me that he and Justin considered themselves "existential professors" engaging in a process for the "exploration of design knowledge" in and beyond JSC. To the end of pushing design know-how further, TransHab chief architect Kriss Kennedy applied architecture's concept of a "vernacular"—a spatially localized style with a structural vocabulary—to spaceflight design.[3] The "local" in this vernacular is not a particular place but environmental extremity in all its forms. The vernacular's "basic vocabulary" elements are transportation, habitats, transitional structures like airlocks, and all the built systems that connect them.[4]

In light of its material and semiotic vernacular, spacecraft and space habitat design evinces discursive aspirations for transitions from Earth to other environments through experimentation.[5] Anthropologist and political ecologist Arturo Escobar describes a variety of social "transitional discourses" emerging in a "multiplicity" of contemporary sites. In these sites, participants work to oppose individualism, disembodiment, discontinuity, and divisions and to promote interconnectedness and interdependence. Transition discourses, Escobar writes,

> posit radical cultural and institutional transformations, indeed, a transition to an altogether different world. This is variously conceptualized in terms of a paradigm shift, a change of civilizational model, or even the coming of an entirely new era beyond the modern dualist, reductionist, and economic age.[6]

While Escobar is concerned with environmental activist and antiglobalization communities and does not explicitly include technical or governmental experts in the scope of his analysis, he lists scientific "frameworks" like ecology, systems of relations, self-organization and complexity, and, importantly, "design" as dimensions of transition discourses.[7] In a review of anthropological work on design practices as social action, Keith Murphy offers an ethnographically oriented definition of "design" as creative and morally "entangled" intentional action that involves "provisioning for one another the conditions of life."[8] But, as anthropologist David Valentine also argues, spaceflight discourses and activities, such as spacecraft design and off-world planning, cannot just embed in the same kinds of earthly conditions, like breathable air and gravity, that invisibly contextualize both human social activities and social theories.[9] Instead, spaceflight design is an elite provisioning practice for select human–machine ensembles in radically transenvironmental spaces. It aims to extend collective living beyond given spatial conditions through the making of transportable and transitional space habitation.

To flesh out the coupled systems theorizations and practices of transhabitation, I attend here to the relationship between the JSC-local and broader transitional discourses and practices of space habitat design. As a discursively powered form of *transitional practice,* space transhabitation work operates within but also outside U.S. governmental practice domains. Peder Anker argues that midcentury government-funded space habitat engineering work became a model for "ecocolonization" aims; closed-loop space colonization designs justified calls for "rational" imperialistic and antihumanistic environmental management of earthly spaces as parts of a larger ecologically closed planetary system.[10] Anker's critique effectively highlights the ways in which environmental power operates in government engineering projects and in NGOs like Engineers Without Borders. But the post–Cold War social variations of ecological and environmental transition discourses suggest, as Escobar's analysis indicates, that ecological systems design cannot be reduced to a monolithic antihumanism or to industrial engineering.

Half of my interlocutors who were working on spacecraft design identified as designers and architects rather than as engineers, and most expressly oriented themselves against the dehumanizing dimensions of industrial engineering. Their work highlights the new dimensions of Western humanism—notably systems-based humanism. Most of these people worked in private practices or in company or university settings, or they were in training for broader industry work—making space habitat design a socially transitive activity. The extent to which space habitat design also happens *beyond* NASA work spaces is evident in the proprietary status of the images produced. Most advanced-concept images belong to individual designers and artists or to government-contractor companies such as SAIC, Lockheed Martin, and Bigelow Aerospace. Following such extramural work took me into JSC's Houston surrounds, to NASA Ames in Silicon Valley, and to aerospace conferences, where space habitat designers link their work to broader professional problems. Functioning as idealized built environments, space habitats traverse U.S. design landscapes as exceptional but charismatic design forms. They are materially connected to futuristic city and community redesign projects as well as private building projects such as sustainable office buildings. Justin and Phillip, like other professionals I met at JSC, had their own businesses and worked beyond JSC's world of insecure funding. As Justin said wryly, "It's nice to see *something* get built, even if not here."

Despite spaceflight design's connections to broader building processes, much of Constellation's Earth-to-space transition-building work happened inside institutional spaces separated from publics and from most of the everyday architectural landscape. Those institutional conditions foster such work but also highlight its dependence on technocratic economies. Like spacesuit design, habitability work took place on the insides of standard institutional buildings, most of which did not exemplify spaceflight engineering concerns with the environmental safety and sustainability of human–habitat intimacies across the solar system. Although the official book on JSC's built history as an institution is titled *Suddenly, Tomorrow Came . . .* , JSC's core buildings are mid-twentieth-century constructs, not "tomorrow's" environmentally efficient and responsive smart buildings.[11] Interlocutors were very conscious of these situated contradictions. Constance Adams has noted, "What was, and still is, breathtaking to me about the facilities that we work in at JSC is how basic they are, yet the absolute edge of the envelope is represented by this place."[12] However, interlocutors perceived their institutional work as productively transitional within a national geography that one Rhode Island School of Design graduate intern described to me as "95 percent full" of "bad, stupid houses."

My fieldwork sites abounded with such morally and politically charged references to the failures of "traditional" design and technocracy and the need for experimentation with extremity. When I sat with Larry Bell and Olga Bannova, both architects as well as professors who managed the Japanese corporate-funded Sasakawa International Center for Space Architecture at the University of Houston, they conveyed the value of opening up design spaces. After handing me the 1991 conference proceedings report from the First International Design for Extreme Environments Assembly, Bell described the need to "transport our minds" into "extreme conditions, environments, analogs" as problem spaces beyond "traditional ways of thinking." For the designers I spoke with and observed, there was a sense of transitionary tension. Their extreme environment habitation experimentation is both supported and constrained by technocratic control—governmental and private—that demarcates the experimentation's social and spatial impact.

Transhabitation, as an aspirational discourse and practice, happens in this in-tension design space that reproduces but also confronts the assumptions and goals of previous closed-loop ecosystem projects. I continue this chapter by locating TransHab as a seminal design project

TransHab miniature mock-up. Photograph by author.

that reflects space architectural engagements with "home" as a processual space in which the conditions of contained dwelling must be accounted for and made transmutable. I then describe lunar habitat conceptual design work trans-forming systemic relations across interiors and exteriors of bodies and machines. I focus on how designers treat that configuration as, in the words of technics theorist Gilbert Simondon, an "associated milieu" rather than as dissociated sets of interior–exterior relations.[13] Finally, I follow translations of transhabitation from JSC outward into terrestrial spaces.

## Transcendental Containers: Cylinders as Political and Ethical Formations

In their most speculative forms, space habitat shapes display the possibilities of design beyond given terrestrial conditions. They often follow the modernist aesthetic of streamlined transport, with sleek contours and spherical shapes. These are also represented as the habitats of future humans and aliens. Built by beings who have overcome situated physical and social constraints to make outer space flourish with life, these livable containers often take diverse forms, like saucers, pods, or, as in the case of the *Alien* movie franchise, exotic technoanatomies.

But down on Earth today, space agency spaceflight forms are driven by the physics of aerodynamic shapes suitable for launch, return, spatial maneuvers, and atmospheric pressurization. Extraterrestrial habitats are being designed according to "realistic" requirements for transportability and immediate usability. This dictates a set of standard forms. At JSC, interlocutors sitting at drafting tables or covered with workshop sawdust often reminded me (sometimes with bitterness) about their constraints. These include the interior dimensions of physics-dictated cylinder-shaped rockets, the exorbitant cost of each pound of transported mass, and the need for extraterrestrial habitat construction using as few rocket launches from Earth as possible. Such constraints pushed to extremes the engineering-inspired modernist dictum that form must follow function, resulting in Apollo-era engineer Max Faget's blunt-nosed habitable ballistic capsule capable of withstanding heated atmospheric reentry and today's orbital space station cylinders.[14] The impact of gravitational physics and propulsion technologies on spaceflight design means that spaceflight habitats are culturally shaped in ways that are familiar but also different from those understood by terrestrially focused anthropologists. The planetary surface habitat shapes I saw on drawing boards and in digital diagrams were cylindrical for existing technical reasons that were not reducible to cultural preferences. But such cylinders become spaces for selectively rearranging cultural forms of life–environment boundaries and relations. Designers and policy makers treat spaceflight's transportable containers as transitional environmental systems-building formations, for advancing design as well as ideology.

Self-contained habitats in the United States have a post–World War II architectural and semiotic history, aligning the emergence of "habitability" as a functional design requirement with broader elite discourses on the social and ecological value of closed-loop habitation experiments in an imperiled world. Habitability, as something more than "survival," started in spaceflight practice as a problem of design beyond the engineering of encapsulated atmospheres. Architects I spoke with and the archives I accessed locate that problem in NASA's early years, when people debated whether space capsules "should" have windows. The story of how windows "humanized" space vehicles was passionately repeated to me also by astronauts and designers, who saw "human-centered design" as vital for astronaut functionality and health as well as for enacting spaceflight's purpose to use spacefarers as situated inhabitants for socially incorporating outer spaces. TransHab architect Adams analogizes the intimacy of space habitat and earthly life: "We're all in this thing together."[15]

Habitability research continued as architects and industrial designers were enlisted during the agency's transition to long-term extraterrestrial habitation projects like building space stations. Bringing non-engineers into NASA was part of an effort to increase the "functional integration" of human, machine, and environment understood to determine long-term mission success. These projects advanced NASA's role in U.S. military and industrial human factors research by focusing on design features geared toward making long-term station tending in hostile environments successful, from the provision of hot water to psychologically correct interior decor. For the design of the cylindrical Skylab, America's first space station, launched in 1973, NASA enlisted industrial designer Raymond Loewy and concept artist Syd Mead, who later designed the "future" for movies like *Tron* and *Blade Runner.* Loewy and Mead were hired to develop so-called man–machine interfaces that would make Skylab's hollowed-out rocket cylinder interior livable. Their work included advocating for a window, designing a better toilet, and apportioning spaces for crew privacy and socializing.[16] Despite these efforts, one JSC architect told me, astronauts reported that Skylab interiors were "ugly as hell." And, like most of NASA's actual vehicles, Skylab bore almost no resemblance to the space stations imagined by futurist mid-century artists and the man known in some NASA circles as the agency's first "architect," Wernher von Braun, whose visions were also translated publicly in the "Man in Space" episode of the Walt Disney Company's *Disneyland* television series and in the Disneyland park's Tomorrowland attraction.[17] Decades later, Walt Disney World's spherical EPCOT (Experimental Prototype Community of Tomorrow) Center was built to showcase a self-enclosed global microcosmic community living in engineered technical and social harmony. But it ended up opening in 1982 as a visitor exhibit.[18] EPCOT was one among a host of 1980s experiments in closed-loop living modeled on NASA plans for the evolution of cylindrical space station modules into more ecologically complex spherical worlds.

NASA's cylindrical BIO-Plex was a successor to its Controlled Ecological Life-Support Systems experimentation program, which occurred alongside other high-profile U.S. government and privately funded closed habitation projects, including Biosphere 2. These ventures flourished as what historian W. Patrick McCray calls "visioneering" projects in response to global North discourses about impending resource crises.[19] Such projects, inspired by cultural authorities like Buckminster Fuller, who popularized the "total Earth system" and "spaceship Earth" concepts, make ecosystems science transferrable to systems engineering; habitats are understood as assemblages of replicable functions, such as thermal, atmospheric, and hydrological systems.[20] From the 1970s through the early 1990s, government military dual-use projects like the cylindrical habitat-based Tektite "man in the sea" program and privately funded projects like Biosphere 2 (Biosphere 1 is the Earth) put humans into machined habitats. Biosphere 2 is not a sphere but a dome-enclosed landscaped habitat in Oracle, Arizona, built with funding from billionaire venture capitalist Ed Bass to be a fully self-regulating closed-loop environment with "crews" of inhabitants. The concept of the "biosphere," which represents the totality of earthly biological life, was popularized in the 1920s by Russian biologist Vladimir Vernadsky, and its cultural significance obtains not just in its spatiality but in its temporality as an evolutionary phase concept. Vernadsky theorized that the biosphere will be transmuted by a "noosphere" of human consciousness and will. In Vernadsky's Soviet cosmist view, quoted in a technical handbook on Soviet closed ecological system construction written by two Russian scientists with the collaborative support of a U.S. NASA Ames scientist, humans are not solely biological beings. As system restructurers, humans are, according to Vernadsky, a "geological force" with capacities to manage a "closed" biosphere that will hold a "freethinking humanity as a single whole."[21] Translating this idea into a U.S. late industrial utopian framework, Bass commissioned Biosphere 2 as an elite-inhabited subbiosphere prototype for closed-loop living experimentation and space colonization planning. Its two "missions" in the early 1990s were complex failures owing to unanticipated ecological and social malfunctions, from insufficient carbon dioxide uptake by plants to disputes among collaborators over costs and procedures. After these controversial habitation experiments, Biosphere 2 became the home of an experimental ecology program run by the University of Arizona. But in its early days it served an ideological function as a vehicle for ecosystemic knowledge making and elite escape-hatch planning.

In 1990, NASA administrator Thomas Paine spoke of Biosphere 2 as an evolutionary systemic unit, naturalizing it as a transitional link in Western colonialist movements out and up:

> In this historical context, I see Biosphere 2 as a shining beacon pointing the way to an expanding future for humanity. Closed ecology systems can free us from Malthusian limitations by making the Solar System our extended home. For the first time in the history of evolution, the human intellect can extend life beyond Earth's biosphere, following the lead of species that left the oceanic biosphere to inhabit dry land billions of years ago. In the twenty-first century, a network of bases throughout the Inner Solar System, interconnected by space transportation and communication infrastructure, can sustain vigorous high-tech civilizations evolving on three worlds.[22]

Paine's "shining beacon" echoes Puritan colonist John Winthrop's vertically romantic 1630 invocation of America as a "city on a hill" to be watched by "the eyes of all people." Paine points to U.S. space programs as a transterrestrial branch in a Darwinian evolutionary tree in which next-stage adaptations take the form not of biological or social structures but of "vigorous" technical infrastructures. The fact that alt-right political figure Steve Bannon, who was later an adviser to President Donald Trump, once served as a consultant on Biosphere 2 is a high-profile example of how such colonial self-containers have served as transitional social forms in libertarian-inflected sea and space settlement ventures. In the late 1980s, President Ronald Reagan invoked the vertical advancement trope of a "shining city on a hill," as he had on other occasions, in a speech announcing authorization of funding for the building of "Space Station Freedom"—which became the cylindrically modular International Space Station. But all of these systems-building projects began in engineering offices and, as I describe here, design studios.

Importantly, like other kinds of spaceflight work, the studio politics of contemporary space habitat design cannot be mapped seamlessly onto policy making or political identity movements. The civil servant and contractor interlocutors with whom I interacted who worked on TransHab and on Constellation planetary surface habitation projects expressed a variety of personal, professional, and political motivations. They were male and female, black, Asian, Latino, and white, ranged in age from nineteen through late middle age, and revealed at least four different religious/areligious affiliations. TransHab in particular represented a professional diversification of NASA's design process—a shift toward interdisciplinary and extramural collaboration that continued into the Constellation era.

As an innovatively inflatable, soft rather than hard, containing cylinder, TransHab was a site of particular forms of transitional environmental politics, principles, and aesthetics that manifested what Keith Murphy describes as design's capacity to materially *and* ethically "enstructure."[23] Although TransHab was not the kind of everyday object that Murphy describes in his analysis of how Swedish social democratic values are designed into home products, it was an object designed to make outer space an improved everyday space and, eventually, a home. In writings promoting TransHab's design values after the project ended, architect

Kriss Kennedy advised design teams to be "technically honest," to "build a constituency of supporters—team, supervisors, crew, management, Congress," to "shield" the team from "politics," to remember that "form follows function." Claiming that TransHab "put the 'living' into 'living and working in space,'" he lauded how "the integrated effort by which this spacecraft was conceived and developed has proven its virtue in meeting tremendous challenges by combining innovative design with cutting-edge technologies that are appropriate for space and planetary surface habitats, with multiple applications for both Earth and beyond."[24] He described the inflatable fabric cylinder as a "single reconfigurable habitat system" created through a "spiral" engineering process that allowed the project to "mature" and become a "widely known" NASA design "brand."[25] TransHab colleague Phillip told me that TransHab deviated from a "hard-shell" engineering heritage, representing a "next generation" of flexible habitat design. Like the creators of the fabric-based BioSuit discussed in chapter 3, TransHab's designers chose materials that facilitated partial separations. These could resist meteoroid and space debris impact but also allow the habitat to be transfigured according to the changing needs of crew life in transit or on a planetary surface.

According to a space architecture survey book made up of contributions by participants in the 2002 Millennium Charter, a meeting of space architects at the Houston World Space Congress, TransHab's fabric-based shape-shifting cylinder belongs to an ancient genealogy of "nomadic" transportable habitats that enable migration and environmental adaptation. TransHab, the book's editors write, keeps "robust, tensile fabric structures with dynamic, bold, sweeping shapes" at the "revolutionary forefront of modern architecture."[26] Although the essays in this book do not explicitly address how their futuristic valorization of nomadism and its materials intersects with Gilles Deleuze and Félix Guattari's nomadic theory, there are continued partial theoretical intersections, as I show in what follows, between postmodern and space habitat design concerns.[27] Phillip interpreted his work in space habitation as a way of "merging aesthetics and function," where both celebrate mobility and transformability as properties of human spatial adaptation capacities through the basic problems of homemaking. Despite the drag of designing within space agency constraints, Phillip was excited about working on what he and others called "basic" shapes and problems, saying, "I see space architecture as renaissance, rebirth—go back to the basics and start with beginnings at [the] core of it."

In spaceflight communities, habitat shapes and ideologies have interrelatable temporal phase typologies. A NASA Ames Research Center "settlement" website organizes NASA's sponsorship of space habitation art in the 1970s according to developmental shapes: cylinders, toroidal ring worlds, and spheres.[28] At JSC, Phillip and Justin described habitation design to me as a multiphase process in which types of designs are vetted and selected, and I was able, as I describe later, to observe the first phase of a lunar habitat design project, which began with shape-type selections. Beyond the design studio, JSC interlocutors described how spaceflight colleagues could be ideologically typologized according to markers like their affinities for kinds of science fiction and their visions of spaceflight socialities. Anthropologist Constance Penley provides a rich description of the reciprocal ideological and aesthetic influence of *Star Trek* and NASA, but within spaceflight communities there are additional schemas of sci-fi-inspired relationships between ideologies and how to live in space.[29] "New Space" entrepreneur Rick Tumlinson describes spaceflight "disciples" as belonging to visionary allegiance groups: "von Braunians" (interested in developing a centralized technocratic and militarized infrastructure), "Saganites" (environmentally concerned scientists who want to explore with robots), and "O'Neillians" (people aspiring toward decentralized space colonization and resource development projects for large populations).[30] These allegiances reflect different visions of systemic and structural living space, with an intimate-to-infinite scalarity.

At stake in these visions of space living, and echoing NASA administrator Paine's aim of turning the solar system into a national domestic sphere, is the problem of how to maintain but transform the concept of "home." Surfacing the *oikos* (home) root of *ecology*, closed-loop ecological systems habitation work has historically paralleled the genesis of astrobiological definitions of habitability as a conditional space/time zone in which planetary life evolves. Both projects converge with midcentury collaborations between astronomers and life scientists, including NASA aerospace life scientists, on the problem of how to think about life-supportive spaces in ways that scale from Earth into the solar system and beyond. These collaborations used terms like *biosphere, ecosystem,* and *ecosphere,* neologisms of the time, coined by scientists trying to label organic/inorganic systemic units and totalities that included humans as, in the words of ecosphere popularizer Lamont C. Cole in 1958, "physiological" parts of an ecosphere energized by the sun, moon, and stars. Such terms became disciplinary boundary concepts that made "life support" scalable—and, as I have argued in past chapters, metasystemic and ecobiopolitical.[31] During Constellation, NASA turned habitability into a "modular" learning program in its Living and Working in Space series of books and other materials for elementary-level science education. A student "engagement" activity in the series called "The Sealed Room" included "content" areas titled "Life in a Sealed Room," "Healthy Choices," "Air and Water," and "Trash or Treasure." The modules were designed to take students through the process of understanding the Earth and the International Space Station as semiclosed-loop systems, the latter relying on resupply and both relying on recycling systems. In this way, the modules universalized "home" as a set of transforming, scalar conditions.

## Universal Homes: Transforming Living Space

During the Constellation program, managers of NASA's Habitability and Human Factors Division carved out a JSC work space called the Habitability Design Center. According to one founding manager, the HDC was meant to formalize the "human-centered" interiors of space habitats, from flight capsules to cylindrical surface habitats being sketched out in "pill," "soda can," "tuna can," "hamburger," and "egg" shapes. In this conceptual space, vehicles were fixed containers that needed to be transformed into "home" as the inherently human-centered basic unit of living. Some architects associated with the HDC wanted to "translate" uncomfortable vehicles into human-centered homes; JSC architect manager Claire told me: "I'm a nester, home oriented, and I've translated that to making home wherever it is, the best home it can be. People have laughingly accused me that I'll be the first person to put a grapevine wreath on orbit." Even if humans were not being truly centered the way the designers wanted, they were responsible for human-centering "habitat living" elements like interior volume size standards and the design of interior accoutrements responsive to the affordances of microgravity and spacecraft environmental control standards. As a project funded

An internal NASA workforce communications graphic from the Bush-era Vision for Space Exploration (which authorized the Constellation program), posted in a JSC office in 2006. Photograph by author.

by NASA, the HDC's experimental work belongs to the history of U.S. government housing design, from mass-produced public housing projects (to which outer space habitats seem utterly opposed) to military outposts (to which space habitats are related). All are regulated forms that materialize technocratic trajectories of social engineering through environmental reengineering.[32] Space architects and designers push this trajectory their way, using the problems of housing in extreme environmental conditions as a platform from which to critique NASA's aeronautical anti-homeyness as well as normative Earth-based design. This work connects with decades of design interest in the transformative and aesthetic potentials of systems thinking as a transition toward universally better forms of living.

Hanging out in the HDC studio-office, I heard vigorous irony-tinged critiques of NASA as an institution as well as expressions of deep affection for it as a conceptually innovative organization. The studio was located at the end of a long hallway of offices peopled by thermal and propulsion systems engineers and human factors experts working on projects like revising the "Man-Systems Integration Standards" manual. Inside the studio, the unapologetic presence of quirky and refined speculative outer space aesthetics deviated from JSC office decor norms, making the studio look like a mixed-media installation. On desks and tables were piles of lunar habitat drawings made with computer-aided design software, next to books on anthropometry and spacecraft engineering. An astronaut Barbie doll and some sci-fi figurines were propped on shelves, and bulletin boards on walls were layered with renderings of fantastical techno-biomimetic things. A portrait of *Star Trek* captain Jean-Luc Picard stuck underneath workaday spacecraft design sketches anchored a wistful sense of what space exploration leadership could look like. Glossy books on design heroes like Loewy and Mead stood near dull-white NASA procedure binders. As an alternative institutional space, the studio was also a manifestation of agency aims to attract and accommodate young workers, creating a space for transgenerational spaceflight design debate. While I was fieldworking in this branch, HDC workers and projects were being featured in architecture blogs and design journals. Sitting on a stool at the Center, I could hear the sounds of their handiwork as they used building tools and materials. Pounding and the noise of sawing came through the walls as they constructed mock-ups of crew capsule and habitat shape and volume configurations. Making full-scale mock-ups was a task the designers had been hired specifically to do, and they loved it. I got oriented to some of these mock-ups and went inside them. Sitting in facsimile crew chairs, I took in the smells of freshly cut materials and the flimsy flexibility of interior volumes being constantly adjusted. There was an ironic juxtaposition going on between the constrained space of the capsules and all the potential living space in the universe.

Defining human-centered design within these constraints was part of the "messiness" that Justin named back in BIO-Plex; a messiness defined by the containment of interrelated systems, including living systems. Sorting out the systemic mess, for JSC designers, involved advocacy for the human-as-system in order to keep it front and center in engineering processes. As in the case of space life sciences work, this was fraught and contradictory. On one hand, designers championed the human system idea, but on the other they argued that design should "drive" systems engineering processes.[33] One soon-to-retire white female engineer branch manager described the problem as a difference between engineering and design ideas about the human system:

> I don't think you can reduce the human to a system in many ways, not in the way that these engineers do that. The typical engineer here thinks a system is, electrons will go through a wire pretty consistently in a black-and-white way, and fluid will go through a tube. Humans are not predictable . . . you have to account for human variability and unpredictability. . . . We advocate for the gray areas here.

While I was there, an HDC-affiliated Danish designer working on mock-ups in Building 29 collaborated with an HDC worker to advocate for the human to be considered a design "subsystem." This argument considers the entire habitat to be full of vitally enjoined systems.[34] But in such views the human-as-system cannot be easily quantified or centered. Once again, it falls into the conceptual spatial gaps that anthropologist of place making Kathleen Stewart calls "ideas as ideals," rich with maneuver room but also haunted by fears that needs and desires might stay "just talk."[35] In JSC design work, the human system is a metrically defined space-requiring organism but also a Western liberal subject with, as one young white male HDC intern told me, a set of "diverse" properties and needs. People work with the idea that such needs can be anticipated through expert knowledge and verified through the use of astronauts as embodied test-case advisers. Donna Haraway argues that technical interventions into life processes make bodies into "biotechnical" and semiotic "strategic systems" that in turn make up meaning-producing fields, to which closed-loop system habitability work, as a conceptual field, belongs just as much as biotechnology.[36]

HDC and other JSC architects and industrial designers worked on metrics-driven projects but also pursued living-space design problems related to their professional identities as culturally-minded nonengineers. In their work spaces it was possible to touch things and be enclosed within spaces that communicated a tangible vision of spaceflight as an actively transformational design space. Architects and designers told me that this kind of vision was *not* exemplified by the International Space Station, which one middle-aged white male team manager described as the product of "engineers" without architecture's commitment to an "integrated project" designed around "how people live . . . what their expectations are." He shook his head and voiced his dream of how better-designed space habitats could effect "a transition" to better space science and living:

> Every astronaut on ISS [is] dealing with bad design all day long. And there's no time to do science, which is what the taxpayer thinks we're doing up there. . . . NASA was born out of the can-do test pilot spirit, and that's a wonderful thing, and it's our heritage, but I think there needs to also be a transition towards these aspects of living in space.

The reconfiguring of mock-ups, distributions of hard-copy and digital renderings and re-renderings, and the alternating flows of visionary enthusiasm and workaday frustration provided a sensual connection to the production of space habitability as a potentially transformational dimension of "good design" for living—even if the technical tasks at hand were as theoretically unspectacular as developing habitability "volume metrics" or as effectually limited as designing furniture and dishware.

HDC work to calculate volume metrics—how much space crew members need to live and carry out their work—was in high gear while

I was at JSC, revealing the micropolitics of volume as a form of governable space. In his work on states' securing of deeply volumetric rather than flat territorial spaces, geographer Stuart Elden calls for attention to the political "geo-metrics" of large three-dimensional spaces like tunnels and the macro-scalar zones of "earth processes."[37]

At NASA, such projects achieve extreme volumetric scalarity by starting with bodily space. The biblical commandment–style requirements language of the "Man-Systems Integration Standards" manual dictates: "Sufficient total habitable volume *shall* be provided to accommodate the full range of required mission functions."[38] According to the manual, bodies on missions exist in a "body motion envelope" that is a "conceptual surface which just encloses the extreme body motion of an activity."[39] The goal of creating a formula for calculating habitable volume would specify the space needed to "accommodate the body motion envelope of the 95th percentile male crewmember performing various IVA [intra-vehicle activity] activities" and account for "social design" considerations like "visual privacy," the spatial privilege of "leadership role," and "proxemics," which "encompasses the study of space as a communications medium."[40] This mind-bogglingly calculative perspective rationalizes Western social positioning values, from body spatial separation to positional equality, in ways that can be formalized and regulated. For example:

1. When conversational or recreational space is necessary, the space should be configured so that the crewmembers can be at distances of 0.5 to 1.2 meters (1.5 to 4.0 feet) and at angles of approximately 90 to 180 degrees from each other. In general, 90 degrees is preferred for casual conversation while 180 degrees is for competitive games or negotiations.
2. Equal relative heights among social conversants should be maintained through spatial configuration and the placement of restraints.[41]

A way of making universal measurements based on idealized regulatory bodies, like the king's foot or the base ten of human digits, the "95th percentile male" body creates a standard for microsocial volume. Such work ties NASA work to automotive design, and I discovered that JSC human factors and design professionals were producing papers for the Society of Automotive Engineers, which supports work related to transportation. During fieldwork I stepped into a JSC 3-D anthropometry scanner once myself and watched as I was made a calculable volumetric shape by powerful new human trait- and type-sorting technology. Nevertheless, HDC designers had faith that such volumetrics projects would open pathways to innovation. Chris, a young white male graduate of the Rhode Island School of Design, spoke to me from across his drafting table one day with a pencil stuck in his mouth, telling me he was hopeful about work on "real-world problems such as how to create a space that is perceived as being very wide and having a lot of volume, and in reality it's actually an idealized volume, really tiny. Because that's gonna make design history right there."

The HDC designers eventually came up with a quantified volumetric formula they valued as part of "good," "integrated," and "minimal" habitability work, but they and other JSC designers continued to yearn for the qualitatively "cool" and "crazy weird"—values they located in what they called "transformer stuff," from toy anime objects to visionary habitats. Over months, I was able to interview HDC workers in a group about their objectives and to accompany them into other meeting spaces. One day at the HDC, a white male engineering supervisor, Randy, who had been out of school for years and had worked as a Lockheed Martin contractor, told me that the HDC existed to go beyond the "zeros and ones" of human factors work and the fallacy of "reenacting the future" by repeating Apollo-era designs. From his desk across the room, Chris, an accomplished artist as well as designer, added that it was a challenge to remain connected to the science fiction art and movies that still inspired him with their "idealized fantasies" of "cool and futuristic" space living, from "virtual reality pods" to "crazy wallpaper." Our conversations would often shift to the possibilities that space habitat design presented, including strategic minimalism and biomimesis. A young European male HDC intern told me, as he worked on a habitat mock-up inside BIO-Plex, that this was where weight and durability constraints meant he could "keep [design] as simple as possible" and regard normal earthly design as luxuriously "extreme" in relation to the efficiency-driving constraints of spaceflight design. But spaceflight design could be extravagant in its own way.

One day I went with all the JSC architects and designers to watch a presentation by Italian-born architect Guy Trotti, who had been funded by the NASA Institute for Advanced Concepts to design a biomimetic-inspired mobile lunar exploration vehicle with scorpion-like flexible transformable features. Trotti was a BioSuit collaborator, as noted in chapter 3, working in affiliation with MIT as well as the Rhode Island School of Design. The JSC participants looked both impressed and chagrined by his unrestrained vision and sumptuous presentation, knowing, as Phillip once told me, that the "crazy" but "great" ideas that Trotti and others had for "transformer robot"–style lunar landers that could "morph" into habitats would not happen "in our lifetimes." Back in the HDC weeks later, George, an older white senior architect contractor with a home business, reminded everyone around the table that while NASA engineering design was often "reactionary," it was important to remember that HDC designers were responsible for giving astronauts "habitability" in terms of the "standard homes" they were used to, not the "zippiest crazy science fiction future thing."

Even when working to "standardize" the dimensions of analogously homelike design, space architects and designers I met in and outside JSC had formulated critical theorizations of what it was like to work at the intersections of architectural theory, systems thinking, and creative commitments to design innovation. In the 1960s, U.S. architects and artists embraced systems thinking as a social salvational mode, visible in systems aesthetics developed by architects like Jack Burnham, but later such thinking was critiqued by artists and architects based on its association with military and industrial technocracy.[42] However, the human–systems integration ideal pursued by JSC designers is a conceptually hybrid systemic concept, sustained by Enlightenment-era as well as modernist, postmodern, and environmentalist ideas. A middle-aged female industrial designer, a former JSC TransHab contractor who now works in a private practice in New York, offered me a transitional discursive critique of classical economics as a form of systemic exceptionalism:

> So, [the] whole idea of looking at a project holistically, it sounds New Age and it's been overused, but I now use the word *systems engineering,* because having read my Smith and my Mill and my Marx and my Hegel, and my Durkheim and my Hannah Arendt and my Foucault, [I know] that economics, actually economy, is not a separate thing from ecology. It's all the exchange of energy. . . . What I've learned in spacecraft design is, if you optimize one system, chances are you're creating problems for the rest of the

> spacecraft. Every system has its own rules, but those rules only function fully in internal interactions. External interactions have to follow different rules, and that involves cooperation with the systems with which they're interacting. Economics in considering itself an independent entity is destroying a lot of the systems it's touching, because of that break, that fallacy in thought.

While JSC designers worked on defining "integrations" and "connections," they facilitated transversal exchanges of Earth-situated and space-situated design concepts in order to give universal scalarity to their research objects—particularly the persistent ordering concept of home.

Phillip, Justin, and HDC designers all mentioned the influence on their work of "Le Corbu" (French modernist architect Charles-Édouard Jeanneret, known as Le Corbusier), famous for his universalizing statement that "homes are machines for living." In applying Le Corbusier's thinking, space architects make "standard" ideas of home theoretically transitory. TransHab architect Constance Adams, as well as people I interviewed at JSC, eloquently collapsed space transportation, home, and living systems metaphors:

> A spaceship is a delicately balanced environment. Every element that doesn't contribute to the overall functioning of the system is by definition working against it. The same is true of our planet—our mothership—though sadly, most architecture today is slowly fatal to nature's systems.[43]

An architect in the TransHab network, German space architect Andreas Vogler, has reworked Le Corbusier's metaphor, calling space stations "complete machines for living."[44] Vogler argues that space stations are "prototypes for universal homes" that can lead "trends" for homes to be autonomously "minimal" in space and energy use in a way that makes them an "active part of the planetary ecosystem."[45] He calls this true "space-age housing," pointing to a persistent horizon of terrestrial expectation unfulfilled by the twentieth-century futurist aesthetic or by human spaceflight.[46] In this view, common to my JSC interlocutors' critiques and to space architecture literature, earthly homes should be made to seem uncanny and unhomely in order to make space "an ecology for living in."[47]

The HDC had its own set of ecological redesign icons, including "crazy," "cool" artists like Joe Davis, with his machine–animal hybrid devices, and the 1960s design collective Archigram. That midcentury avant-garde group is known for its experimentations in survival technology featuring reconfigurable modular "plug-in" and livable "pod" concepts. Archigram members' avowedly apolitical quest for small-scale transportable design, following Buckminster Fuller's formulation of "ephemeralism" as the value of doing more with less, was inspired in part by NASA lunar modules.[48] Deeply influenced by systems theory and cybernetics, the Archigram collective published editorials, such as an essay titled "Open Ends" celebrating nomadism, metamorphosis, and modern systemic failures as opportunities to get back to the "basis" of design, such as "the exchange of facility between one object and another."[49] This concept posits that the properties of human, machine, and environment can be made exchangeable to "increase the possible means of our survival."[50]

At JSC, systems engineering and avant-garde industrial design dovetailed in the rethinking of the relational boundaries between bodies, machines, matter, and environments. Architects and designers told me they felt they were uniquely capable of doing this; Phillip, a TransHab veteran, expressed it as a form of exemplary systems thinking:

> For the folks that go through engineering school, the mentality is all about optimizing the design. . . . And that's not always the right solution, because it needs to be an integrated solution that balances both psychological and physical environments so that you give them the best environment in which they have to live and work. . . . We [architects and designers] make pretty good systems engineers because we're not trained as a systems engineer.

## Systemic Exchange: Associating Environments

The space station systems integration engineer who told me, "Our thought process at NASA is . . . environment, environment, environment . . . everything interacts with everything" (as quoted in my Introduction) was, when we met, involved with readying the deployment of water filtration devices in the station so that crews could drink their own urine. While I was at JSC a media call went out to the Houston area for experimental volunteers—come pee in a bag and be a part of NASA's water recycling research. An older white male water systems engineer involved in the project showed me working prototypes for the system, of which he was very proud, which included collections of machinery hooked to containers of murky faux-urine test substances. Waste recycling is a crucial closed-loop living ecological system design problem, fostering processes that equate and transpose systemic matter in terms of its conditional functions and features. Urine = water. In this systems design space, hard notions of body integrity and material categories loosen up. They become reconfigurable elements within a domain of flow schematics, trade spaces, and efficiency innovations. In space habitat design, the body and its emanations are structurally and functionally distributable, making the cyborg human-machine concept seem retrograde in its consolidated hybridity. In addition, spaceflight environments such as habitat interiors and planetary spaces are integrated (or "balanced," as architect Phillip says) by being treated as connectable and/or separable. In this design space, waste is both vitally toxic and systemically constitutive—not just in a material organic sense but also in technical theoretical terms.

As I sat for months in Building 29 watching "lunar habitat brainstorm" sessions, the "waste" category conjoined humans, machines, and extraterrestrial spaces as things with contingently interacting environmental interiors and exteriors. This was more crazy-weird work in cultural terms, but rather matter-of-fact in environmental systems technical terms. The technology theorist who provides the most resonant conceptualization of what goes on in this kind of design space is Georges Canguilhem's student Gilbert Simondon, whose treatment of "technical objects" as emergent entities provides machines with internal and external milieus akin to those of organisms. Simondon theorizes object-emergent processes as systemic and spaces as environmental, which leads to his concern with the nature of machinic milieus. This concern is informed by empirical attention to what Simondon, a philosopher who also actively engaged in tinkering with things, calls "technical beings" in states of "individuation." Simondon and his students based his work on opening up machines during the time before machine innards were made opaque to a world of consumers conditioned to make serial purchases and discards. Working at the edges of cybernetics and autopoietic theory, Simondon describes how "technical beings" emerge from relations between objects and the simultaneously "natural and technical environment" around them:

> It is a certain regime of natural elements surrounding the technical being, linked to a certain regime of elements that constitute the technical being. The associated milieu mediates the relation between technical, fabricated elements and natural elements, at the heart of which the technical being functions.[51]

Simondon's theorization of how the explicitly systemic and cofunctioning properties of technically defined objects and environments constitute an "associated environment" is visible in spaceflight work with spacecraft and external environmental ensembles, in which crews and their biological (natural) elements are technically incorporated as parts of systems. This illuminates how waste is dealt with by systems-oriented designers who consider human, machinic, and natural things to be internally and externally environmental for one another. Philosopher Anne Sauvagnargues notes that Simondon's transductive theorization of a technical individual "in progress" makes interiority "relative" and interior/exterior separations (like those we have seen with spacesuits) more "membranous" than closed.[52] This shows up in contemporary engineering projects that use human movement and waste to power machines, air vents to cool computer servers, computer heat to warm buildings, and machines to fix carbon dioxide out of atmospheres.

In transhabitation work, thinking in terms of an associated environment makes toxicity into an innovation driver and human waste into protective material. Traveling through Houston to meet with contractor architects who worked on TransHab and other current projects like lunar habitat design, I ended up one hot, windy day at the Galveston studio of a middle-aged white male architect who designed nontoxic tools and utensils for TransHab and the International Space Station. I held some of these plain plastic molded things, marveling at their emergence from work on with toxicity and function in a closed ecological space. Like other architects in charge of designing things for a "human-rated" (versus robotic) spacecraft, Gil had to understand materials science on a molecular level to know how much any given material would "gas off" into the craft interior. The interior of a spacecraft is a contained atmosphere that is precisely designed and monitored to maintain a livable version of what anthropologist Nicholas Shapiro calls the "chemosphere" of modern- (terrestrial-) built environments.[53] Shapiro examines how people's fine-tuned embodied responses to the pervasive home-based exposure to late industrial "material ecologies" create a "chemical sublime" experience that inverts the Enlightenment idea of the sublime as a mode of reorienting minds to transcendental reason and capacities.[54] At JSC, Dan, a colleague of Gil and a "behavioral toxicologist" in the Habitability and Human Factors Division, once complained to me that "you need more of a philosophy degree than an engineering degree to write requirements." He searched for ways to convey the "functional philosophy" of "clean air" in requirements statements, rather than just specifying that air samples should not exceed "2 ppm [parts per million] benzene"; he worried that the latter could prevent engineering "ingenuity and creativity" in an era of increasingly "integrated" human and machine systems in which things like air, breath, and gases could be worked with in new ways. The desire to keep spacecraft design a "philosophically" open-ended relation-making process, even in the face of danger, fosters the sense of a systemic sublime. Design becomes a way to reason in terms of systemic function rather than in terms of essentialized things, a process I saw manifest in the treatment of waste in lunar habitat design.

The Constellation lunar habitat brainstorming sessions I observed at JSC were held in a small room near BIO-Plex, which designers called the "war room." Phillip, Justin, and other manager-level designers ran the meeting as a studio space for developing habitat design options, some of which came from past experience with the short-lived Bush Sr.–era lunar return program. The room was often full of computer-aided design drawings from previous meetings; also visible was a poster reminding participants of NASA meeting ethics: "Listen respectfully." Meetings generally included from five to ten people around a table and others attending via conference phone from other sites and centers. Participants represented a variety of disciplines, including architecture, industrial design, engineering subspecialties, flight medicine, and mission operations. The customers for the team's finalized design concepts were decision makers in Constellation's Lunar Architecture project; as I explained in my Introduction, *architecture* in this usage refers to the overarching assembly of mission technologies and operations plans. Mission operations were being debated during this time, meaning that it was not clear what astronauts would be doing in their habitat on the moon. The senior architects felt drawn in, as they often were, to designing *to* vehicle specifications rather than *for* the purposes of missions. However, Phillip and Justin were also training the next generation of architects and designers for that agency environmental constraint. Over months, the process was overseen by middle-aged male team leaders, but male and female student interns from local universities also contributed, and their ideas were treated respectfully. Ultimately, the team's habitat designs would be "down-selected" within a design "trade space," in which quantitative representations of the building materials and processes would be compared and selected "in" or "out" of the design in terms of system-relevant characteristics, from weight and risk to cost and functionality. Using drawings and spreadsheets, people discussed how to design for engineering requirements but also how to innovate.

In several brainstorming sessions, the problem of how to shield inhabitants from the space "radiation environment" put waste into relation with habitat structuration. As if paralleling Simondon's and Deleuze's points that it is at the exteriorly connected edges of things that philosophical ideas about the existence of a wholly separable interiority lose salience, habitat design became officially "semiclosed" but conceptually semiopen, providing an opportunity to control associations among bodily interiors, habitat interiors, and planetary environments.[55] A student intern, Susan, came to the brainstorm table one day with an idea about how to transform waste into building material, an idea that engineer and theorist Lydia Kallipoliti puts into the context of twentieth-century aerospace and counterculture closed-loop ecology habitat design, in which the potential of worlds without waste and loss created both a paradox and creative space for design.[56] Design products had to be inherently retransformable in spaces imagined as encircled against extreme outer environments. The transhabitation design processes I witnessed were just as ecologically concerned, but not fully closed. They were based on expectations for careful "in situ resource utilization" practices that would bring outer space terrain into the habitat's larger associated environment.

Susan brainstormed from her intuitively naive position as a student: "We can compact trash and build a [radiation] shield for the habitat, but we need to know how to calculate the production of trash." A trade space of associated environmental elements opened up. Trash was one habitat-shielding option; another would be to cover the habitat with lunar regolith (that is, lunar soil). A senior engineer countered that yet another option would be to "stick with water," which would be a relatively easy-to-handle "systemic" renewable resource, unlike the current standard material for shielding spacecraft interiors from radiation, polyethylene, which is a "single-use chemical." Susan agreed that "water" could be "scavenged" from "the system," where "water" to all at the table meant potable water, gray water, and urine. As space "astrotect" Marc Cohen and his colleagues have noted, water is a "massively redundant" passive building feature that is both usable and consumable.[57] "Water" in the brainstorm conversation was imagined cyclically and in multiple forms, which suggested the outlines of another nested trade space that included human alimentary systems as building material processers: urine or trash could be used as shielding, therefore radiation shielding could be as simple and light on launch as deflated bags ready to be filled with

A space architect's table in the Habitability Design Center showing a pill-shaped mock-up design. Photograph by author.

trash or urine. An animated discussion ensued: transiting astronauts would be protected by water and their own urine, both of which could be used to make protective sleeping bags in which astronauts could "ride out" a galactic storm of cosmic rays. These images caused laughter around the table but evoked the image of what one skeptical NASA engineer I met had dubbed, with rolling eyes, the "econaut," a hyper-ecologically aware and integrated astronautical human. Andreas Vogler and his colleague Arturo Vittori address the treatment of waste as a hallmark of space architecture's distinction as a "new field." They state that "on a spaceship the concept of waste is redundant," meaning that "waste" exceeds its systemic specificity as "resources in different states of processing (just like in a natural system)," but they also invoke the famous NASA value that things must have multiple uses.[58]

During this brainstorming session, the Western moral taboo against living "in" one's own waste was not mentioned, because in the habitability trade space waste was transformed into a practical, resourceful, calculable element, redundant in an engineering sense, and evaluable both numerically and intuitively. According to the handbook *Human Spaceflight: Mission Analysis and Design,* a work sponsored by the U.S. Department of Defense and NASA that I saw on many office shelves in the Habitability and Human Factors Division, humans are, "thermodynamically speaking,"

> open systems—they exchange matter and energy with their environment. They consume matter to provide the building blocks for biosynthesis and the fuel and oxidant required to run their biological "engines." The engines produce energy for growth, mobility, and maintenance of internal (human) systems. This "combustion" process produces thermal and chemical by-products. The "matter" people need is mainly food, water, and oxygen, and their main outputs are heat and metabolic products such as sweat, urine, feces, and carbon dioxide.[59]

In brainstorming discussions about the habitat's "semiclosed-loop system," human waste became metaphorically and practically the same as other output and other available matter for radiation shielding, such as lunar regolith: all are polluting elements that can be transformed into shielding, plant fertilizers, or building materials.

The animated conversation quickly turned to a consideration of the ways in which waste would be generated in general, a move that began to materialize, in habitability terms, what human factors experts call the "lateral loop" of human systems. A senior engineer with a lot of space shuttle and space station experience connected this brainstorm problem to a problem faced by another JSC department: "Guys, the food people [JSC dieticians in charge of space shuttle and space station nutrition systems] are trying to minimize trash by having the astronauts eat together rather than all these individual meals with individual packages that all weigh something—you know, family-style meals." Going with that idea, he offered, would save on mass and volume. But then the discussion turned to the fact that family-style meals would mean less trash, and therefore trash would become a less reliable resource for building or protection. One of the architects summed up the state of the trade space at this point, noting that family-style meals "cut down on waste, but increase the social benefit." All of a sudden, both "styles" of eating and radiation protection associated with the internal spatial dynamics of social cohesion became intimately relevant to problems of survivability. Phillip summed up the associated environmental conditions of habitability trade work, saying, "In order to do a fair trade we have to account for other things."

All of these trades in meanings, use values, and assumptions about matter in and out of place were situated within the space architecture visions of the universal house; homes must be understood not just to exist but to exist in association with systems and environments. The idea that a space habitat might serve as a model for a home links to historical Western utopian visions of so-called healthy houses, houses designed to promote perfectly functioning bodies and families, that emerged in the mid-nineteenth century. As architectural historian Victoria Solan argues, health and home became inextricably intertwined from the mid-nineteenth century onward, but conceptualizations of the "healthy home" did not always align with prevailing medical theories of health. Sometimes alternative explanatory models for good dwelling were based on utopian concerns not with cleanliness or reduction of exposure to the elements but with fitness, resilience, and a return of the integration of people with nature.[60] The lunar habitat brainstorming session, with its futuristic ecological proposal of living better "in" toxic or dangerous substances, reframed the "healthy" environment as a semiopen systemic trade space.

The group's attempt to account "fairly" for other things in a systemic yet intuitive way, by mapping trade-off relationships among building materials, survival, waste, and social cohesion, made up the material and moral economy of what veteran JSC architecture expert Phillip called taking a systemically "hab-centric point of view." Ecosystemically shifted off base here is a "centered" human. All habitability elements, living and nonliving as well as inside and outside the habitat, can be separated out and assigned numbers so that upper-level managers can rationally select among them for habitat designs. Preparing solutions that made for what interlocutors called combined engineering and design habitability "goodness" was also a way to organizationally distinguish space architecture expertise in making the "hab-centric" point of view workable, such as technically and humanistically accounting for bodily/social/environmental systemic causes, effects, and relationships. Most of the time during the brainstorming session, the ecological politics of this hab-centric accounting stayed within the cultural economy of the NASA trade space, but once in a while the politics transited into larger social landscapes beyond NASA's interplanetary designscapes.

Like Gil, the designer of nontoxic housewares, a number of the architects were involved in both JSC architecture and the U.S. architecture landscape. Judith, an Arizona-based industrial designer who had worked on TransHab, offered her thoughts on working between Earth and space. I asked her to explain the relationship between her stated interest in the future of earthly infrastructures and her interest in space colonies. She responded that she agreed with other architects, such as Constance Adams, that this all had to do with "big picture" processes, such as helping Earth reproduce itself. But she slipped back easily from viewing the human relationship to the "big picture" as primarily biological to seeing it as primarily technical, as an exercise in what she called our unique ability to merge "*techne* and logos." Judith was interested in working with and within the "big picture" as one big design concept space:

> I've had people call me, oh, so you do biomimesis. . . . I'm not setting out to do biomimesis, and I'm not going to start my next design by looking at a leaf. I'm going to start out my design with the technical challenges, and I'm going to work the various systems. Chances are when I start getting close to a solution, when I'm having only a couple of options now, when I'm trading, the chances

> are there's a paradigm existing in the natural world, at whatever scale, it might be microscopic, it might be macroscopic, it might be cosmic, right, depending what is the problem [to be] solved, that will give me a cue as to what will be the more efficient way to go, what may be the more elegant way to go.

When I asked her if she does this by working to combine two different domains of nature, Earth and space, she replied, "There's only one paradigm, it's all the same." Over the course of the rest of the interview, it became clear that Judith was referring to one existential paradigm for design, for living, and for inhabiting, illuminated by engagements with extremes and the conversion of things into quantifiable systems terms.

While I was at the HDC and in the lunar habitat brainstorming sessions, I got into mock-ups elaborately marked with grids for measuring the distances between people's bodies and the surfaces of walls and other things, which were used in gauging people's perceptions of roominess as a function of where they stood or sat. The grid lines allowed designers to statistically analyze data on "sociokinesis," or group movement patterns, in order to establish the "quantifiable relationship between environmental factors and human behavior."[61] This underscored the fact that space architecture contributes to a powerful modern technocratic orthodoxy of public space study and design, and also that such work can be restricted from public view, as when the internal JSC document explaining the HDC's sociokinesis data was for a while categorized as "sensitive but not classified." Nevertheless, this work to both grid out and transform the nature of habitability elements and then to trade them according to off-the-grid flexible desires and aspirations makes it impossible to reduce transhabitation concept work to the purely technocratically instrumental. Space architects and their concepts also travel outward and engage with other environmental and public projects beyond JSC's gated walls.

## Translations: On the Road

It is possible to follow transhabitation work outward from JSC, into the corporate production of Bigelow Aerospace's space station inflatable habitat module and hotel prototype, dubbed "Son of TransHab" by NASA workers, and into the work of one architect who designed and promoted a small mobile camper trailer based on his work at the HDC. Having garnered awards and media fame making furniture for clients like small-scale-event theorist Malcolm Gladwell and for designing "small houses for the next century," HDC contractor architect Garrett Finney (as an article on his work declares) "brings space habitability down to earth."[62] Finney trained at Yale with an emphasis in blacksmithing and was the 1994 winner of the Rome Prize in Architecture. In his Houston studio, he designed the Cricket pop-up camper trailer not as a "house on wheels" but, following his work on transportable habitats for NASA, as "equipment."[63] Although he does not cite space architects like Andreas Vogler as inspiration, Finney's habitable equipment follows Vogler's "three principles" for space design: it is mobile, autonomous, and interactive.[64] Finney detoured Buckminster Fuller's call for a design revolution into what he called a "camping revolution," and in doing so, he wrote to me once, he was going "Cricket crazy." Cricket was crazy-cool, good, minimal, and all about open systemic integration. It was a response not just to a market for vehicles but also, as Finney told a local newspaper, to "the problems of state and national parks, the ironies of eco-tourism, and designing a twenty-first-century sustainable campground."[65] The Cricket prototype was a transformable and transitional object, translating the universality of spaceflight home-building concepts and practice through the act of driving it out of Houston and into American landscapes.

Finney took Cricket on what he called a "shakedown tour," borrowing the maritime and aerospace practice of doing a final structural test by putting a prototype vessel into motion. With its lightness, its modularity, and its insect-like silhouette, it is a terrestrial concept incarnation of architect Guy Trotti's scorpion-like lunar exploration vehicle—only Cricket exists and works. A *House & Garden* article published while he was working on lunar habitat design called Finney "the Thinker" among a bevy of "New Tastemakers." In the article he summed up the HDC ethos: "I see the world quite differently having worked for NASA. Sustainability is certainly creeping into my work. I am very aware of what civilization takes for granted here on the mother ship."[66] A Houston news article also made sense of the Gladwell connection by explaining that Garrett was a "natural-born" example of what Gladwell calls "connectors," as in people who "make linkages" and "spread ideas."[67] But

Finney was also responding to his environment, Houston, as a place from which to escape, physically and conceptually:

> Why is it so uncommon to take into account the environment you are building in and adapt the building to it? While this seems obvious, it must not be, because it is clearly not done. For example, here in Houston so many houses are built with no consideration of how to keep them naturally cool. Instead, everyone leans heavily on the air conditioner instead of designing to minimize its use.[68]

Finney's airy, flip-open camper was designed to be self-cooling, or, in space analog and biomimetic terms, to have a circulatory system. He demonstrated his interest in systems integration by making space design's habitability concept sensible in a publicly empirical sense. On the road, Cricket could be seen as innovative or just plain weird and not very sensible; it could appear as a new concept or as a rickety-looking trailer thing. But Finney intended Cricket to invite speculation about its relationship to other things and spaces. In photos of Cricket and in diagrams drawn like lunar habitat cutaways, the little camper's ready-to-go trailer hitch reminded me of a quote from John Muir that is often found in environmentalist literature and websites: "When we try to pick out anything by itself, we find it hitched to everything else in the Universe."

This chapter has described the making of space vehicle "habitability" as a transitional discursive process that takes the form of universalizing transhabitation work. In this work, systems design to provision ensembles of people and things for extreme conditions relates transportation and transcendence of interior and exterior environments to Earth and space design. Transhabitation work bridges gaps between technical and ecological theories of habitation and between late modernist architecture and current engineering design practices. While processes to define habitability are meant to get specific things designed at NASA, they also open up design research to spaces in which the systemic relations of inhabited environments are transterrestrial. These processes also aim to make use of and prevent American worst-case scenarios of limitation and contingency—to keep under the governmental roof some grand design challenges to demonstrate ideal forms of human dwelling. While I was watching these processes, a senior manager at NASA lamented a possible future for spaceflight design and building: "I can definitely picture scenarios when NASA will end." However, Chris, the HDC designer, spoke of his own faith in the transcendent value of spaceflight, arguing that it would persist because it "feeds the collective soul." Transhabitation makes visible ways in which U.S. work on environments and systems generates transitional discourses and practices. As I explore in the next chapter, these processes also call into question the home boundaries and exterior relations of Earth as a planet.

# 5

# THE SOLAR ECOSYSTEM
## POSTTERRESTRIAL ECOLOGIES AND ECONOMIES

When scientifically represented as a changing planetary "system," Earth rematerializes into different whole forms of itself. Western geological and biological practices multiply the planet into sequential versions, creating epochal Earths that stand in relation to each other as well as to other things in the solar system. In the beginning of that scientific epochal story, before life changes everything, Earth is a ball of solar systemic material. Life's appearance on this geological orb creates an ontological rift in scientific space-time and opens a dramatic new chapter in the story: a terrestrial lifeworld detaches from lifeless outer space and, in return, extraterrestrial space becomes separately nonearthly. Scientists distinguish the subsequent terrestrial lifeworld epochs according to the emergence of new ecological systems and catastrophic extinctions. Within the geological strata that experts dig into to construct the deep history of a multistoried Earth are particles and chunks of material left by large extraterrestrial objects—comets, asteroids, meteors—which can, and will, foster or destroy life upon impact. Or, as this chapter narrates, these materials can be engaged to remake the material boundaries of ecological and economic space beyond the terrestrial surface and atmosphere.

Asteroids and comets with planet-crossing orbits may never hit a planet, or they can be imminent threats. In free solar systemic space they travel as partial-worldly bodies; if they impact they become postcollision parts of whole geologies. These traveling bodies behave contingently in relation to the gravitational forces and motions of other bodies, and their orbits can keep them near planets or take them to the ends of the solar system. Astronomers refer to comets and asteroids with safe or potentially threatening Earth-crossing orbits as near-Earth objects, or NEOs.[1] Paleo-NEO impacts brought water and other ecology-building substances, maybe even proto-life-forms, to the Earth from beyond, creating new species niches as well as causing annihilations. In a broader conjoined Earth–space system, NEOs and other asteroids and comets are materially betwixt and between. But these borderline attributes are no longer out of bounds. For scientists as well as engineers, policy makers, and venture capitalists, NEOs are systemic materials to work with, conceptually and physically. This work rematerializes relations between contemporary planet Earth and an exogeological solar system.

One way to perceive JSC's location in North America is geological rather than geographic—it is triangulated within NEO impacts. To the north and west, desert lands preserve remnants of the Odessa Meteor Crater and the Sierra Madera astrobleme (star wound). Due south is the Gulf of Mexico, where annual meteor showers slash across night skies above the glaring lights of oil rigs. Hidden under the gulf's deep-seated waters at the edge of the Yucatán Peninsula, but visible in oil industry radar-prospecting imagery, lies the great Chicxulub crater. According to a new geological forensic story, a NEO with a circumference of about six miles impacted there 66 million years ago. This event corresponds with the end of the Cretaceous geological and dinosaur epoch and the beginning of the contemporary Cenozoic "age of mammals." The remains of the dinosaur killer are like an extraterrestrial thumbprint: a layer of iridium-laced material is found planetwide. Iridium is rare on Earth and common in asteroids, and this iridium-laced stratum has come to be known as the K–Pg (Cretaceous–Paleogene) boundary.[2] Sample chunks of what geologists and rock-hound vendors call "K–Pg boundary material" are for sale online. Like other earthly meteoritic objects, NEO matter is loaded with meaning about cosmic systemic connections and separations. For one thing, NEO impacts have conjoined astronomical and terrestrial evolutionary science in the last century. And during the recently defined Anthropocene epoch, characterized by centuries of unevenly distributed industrial impacts, NEOs have shifted from being astronomical objects to being environmental ones to manage and exploit.

Projects to control and empower NEOs, like one I followed at JSC and describe from here on in, remake understandings of Earth's material systematicity by reattaching it to the solar system. As a NASA Headquarters administrator told me emphatically, "We're not waiting for NEOs to hit. They hit *all the time.* People don't realize but the planet gets 80 to 100 metric tons of incoming material every day." Today NEOs are being assigned roles in national environmental security and economic futures. As atmosphere-crossing things, NEOs are akin to the sun or cosmic radiation in space-focused social practices. They can be treated as having properties to be protected against and to be harnessed. The iridium in the K–Pg boundary, for example, is a form of platinum, signaling the material multiplicity of NEOs as things with dangerous ranges and extractable resources.

In this chapter, NEOs have become *boundary material* for practices that give the solar system forms of ecosystematicity and scalarity. NEOs are social theoretical boundary material as well. In keeping with Susan Leigh Star and James Griesemer's original theorization of boundary *objects* as things that disciplines share to consolidate "institutional ecologies" of knowledge production, NEOs align practices of space agencies, militaries, international science networks, and space interest groups.[3] As compounded problems for science and governance, they are, in the words of moral philosopher Martha Nussbaum, natural objects being multiply "objectified" in terms of their "instrumentality," "boundary-integrity," "violability," and "autonomy" in ways that "urge us" to "think differently about parts of the natural environment."[4] But NEOs are also solar systemic parts whose material nearness is marginal but potential. From a new materialist theoretical perspective, NEOs fit within the diverse particulate fabric of a universe not bound to social imaginaries of stable boundaries.[5] But, as anthropologist Stuart McLean argues in his examination of how protean muddy margins confound and undermine European imaginaries about stable landscapes, new materialism theories tend to focus on things in explicitly social—and, I would add, terrestrial—proximity, and can therefore background certain kinds of spaces and materialities.[6] Like mud, NEOs are "interstitial" material that people imagine to be far beyond social and earthly worlds, but in their emerging ecological and economic significance they pull together powerful discourses about earthly limits and end points—and in this case, the possible futures of broadly environmental social worlds.[7]

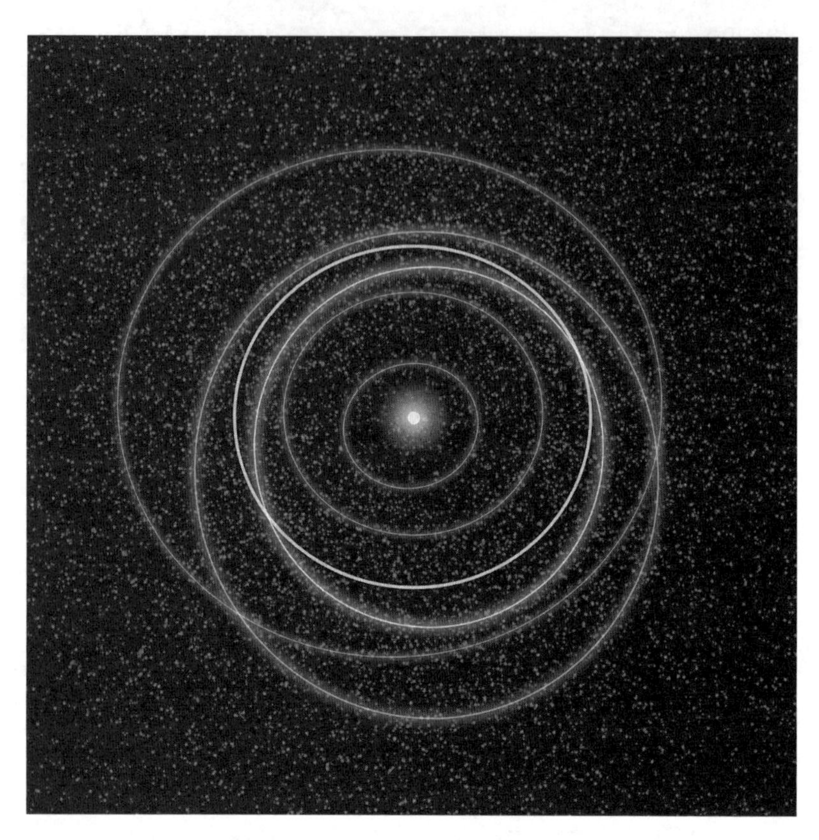

This 2012 image models the known NEO "population" (rendered in dots) among and beyond the solar system's "inner planets": Mercury, Venus, Earth (rendered in bright white), and Mars. It was produced at the Jet Propulsion Laboratory using data from its NEOWISE (NEO Wide-field Infrared Survey Explorer) asteroid position and type detection mission. The image also includes the modeled orbits of two NEOs: one that passes near Earth's orbit and another "potentially hazardous" one that intersects it. Courtesy of NASA/JPL-Caltech.

## Transboundary Material

Social labor to control NEO boundary material and its spatial transitions ties them to modes of thinking differently about both the solar system's material nature and the "human" environment. People—from scientists and policy makers to venture capitalists—working with NEOs in environmental and systems terms turn them into developmental material for postterrestrial ecological and economic systems. Their aims are to keep NEO matter partially separate but also potentially integrable as system-building material. As boundary material, NEOs connect projects of cosmological relationality, wealth making, and security in an integrated solar environment. They shift popular understandings of the solar system, from a space of planets and moons to an environment full of diverse forms of matter.[8] NEOs are also being used to set up new environmental boundaries for social groups included in—and excluded from—relations with outer space as both a threat and a colonial resource. If, as anthropologist Elizabeth Povinelli argues, there is a geoontological politics of relations with nonlife that gets elided by a privileged politics of life, NEOs pull those politics outward beyond the geoterrestrial.[9] They shed light on the scalable processual materiality of systemic thought and practice, in which certain kinds of boundary materials—such as radioactive isotopes, oil, and solar energy—define systems and enable intersystemic relations that exceed everyday social spaces. Nonearthly but relationally geological, NEO boundary materiality illuminates the emerging environmental history and political ecology of a solar *eco*system—where *eco* encompasses ecology and economy.

Contributing to making this ecosystem are experts whose work on asteroid "planetary defense" and exploration systems during the Constellation program ended up influencing Bush-era and Obama-era policies for a businesslike solar system. During my fieldwork I followed a group of JSC interlocutors who were developing a human-to-NEO exploration mission concept and narrating the potentials of human–NEO relations. Their open-ended systemic view of NEO exploration intersected with global-scaled NEO activist projects to track NEOs and fix them into national and international policy structures. In both cases, experts treat human evolution as contingent on humans' ability to control solar systemically scaled space, dynamics, energies, and materials. By tacitly responding to the critical Earth systems science perspective that it will take multiple Earths to maintain U.S.-defined forms of prosperity, NEO activists open a kinetic solar system of material relations and connect spaceflight with the fate of the human species. This is a technoscientific version of what Kathleen Stewart and Susan Harding analyze as forms of affinity-based U.S. apocalyptic and millennial activism spurred by concerns about salvational struggle and sustainability.[10]

With respect to the trajectory of this book, NEOs traverse a solar ecosystem that is being aspirationally constituted through metasystemic modes, ecobiopolitics, connection and separation technologies, and transhabitation. Their presence in social models of space/time situates terrestrial geology within a solar material history. In this chapter I follow how NEOs' boundary materiality manifests in science, policy, and spaceflight advocacy, keeping open, as I have endeavored to do throughout this ethnography, a view into the varieties of spaceflight environmentalism.

## Detecting: Matters of Stability and Security

In summer 2016, I got a message from a planetary defense officer. When I met Timothy at JSC in 2007, he was the RIO (Russian interface officer) flight controller I wrote about in the Introduction, my escort in JSC's International Space Station mission control room. His current workplace, the Planetary Defense Coordination Office in NASA Headquarters, is tasked with supporting the congressional directive to "defend the planet" and the Asteroid Redirect Mission concept for NEO capture and deflection technologies, initiated under the Obama administration. Timothy let me know about two upcoming back-to-back events associated with the Jet Propulsion Laboratory—a meeting on "small body" science and the first-ever asteroid impact disaster response exercise conducted by the Federal Emergency Management Agency (FEMA). Following the 2015 meteor impact disaster in the Russian oblast of Chelyabinsk, the U.S. government decided that NEOs should be treated like other on-the-ground natural disasters. Timothy was selected to work in planetary defense because of his expertise in NEO mission planning; at JSC he was part of a NEO activist group that he and some of his colleagues referred to as "the NEOphiles."[11] The NEOphiles authored and widely promoted a white paper on their study of a human-to-NEO mission, a "paper NASA" mission that others in the agency considered a distraction from Constellation's moon–Mars–beyond planetary trajectories. NEOphile topophilia parallels that of extremophile biologists as a fascination with NEO extremities and evolutionary significances. Through NEOphile stories and work, I was taken beyond the stable planetary spatial system of my childhood. NEOs have been assigned a part in an ongoing transformation of the Western heavens, taking them from profanely natural to manageably environmental. From NEOs' designation as celestial anomalies to their identification as systemic hazards, NEOs have become building blocks in the structuration of extraterrestrial material detection and defense.

The story of the NEOphiles' transcendentally motivated work to move NEOs from policy margins to centers bridges a longer historical tale about the solar system's scientific and political rematerialization. During my time following their program activism, which took the form of formal mission development as well as voluntarist public outreach, the NEOphiles were labeled in media reports as "rebels" and a "small band of believers." I also heard them described at JSC as "heretics" threatening Constellation's lunar-return and Mars-forward architecture. In conversations with NEOphiles, we would play with the "neo" referent, such as linking NEOphile efforts to those of the hero Neo in *The Matrix* movie series, a savior called to rescue humanity from biotechnological enslavement to an inhuman artificial intelligence system. When Timothy and his colleagues travel periodically to the high, cold spaces of Hawaii's Mauna Kea Observatories to do NEO observations, they like playing Blue Öyster Cult's song "(Don't Fear) The Reaper."

Yet NEO activism cannot be reduced to productions of a new "fear ecology"; rather, it is concerned with gaining control of human–solar system relations through NEO defense, deflection, capture, and use.[12] Most of the NEOphiles I followed at JSC were white and male; they had global allies, some from Japan whom I met at JSC, including life scientists, doctors, physicists, astronomers, astronauts, and venture capitalists. The JSC NEOphiles' sense of having a place in a deep species history was crucial to their identity as members of a network with a "calling" and a genealogy of iconoclastic scientific and science fiction forebears. The NEOphiles gave me reports and materials and invited me to talks they gave so that I could learn the historical cosmogony of what Timothy jokingly called the "NEOverse."

By the early twentieth century, solar system materiality had been, in astronomical and geological terms, reconstituted through understandings of planetary relationality. What exactly outer space was made up of had been a matter of debate between vacuists and plenists, but the celestial mechanists of the nineteenth century had jettisoned the Cartesian proposition of a plenum made up of objects embedded in separate vortices set in motion by a deity. The mechanists focused instead on the Newtonian imaginary of a largely vacuum-filled space in which planetary orbits reflected calculable laws of relational force, motion, and order. They were sure that calculative anomalies of planetary orbital spacing in the Copernican solar system model would be reconcilable when "missing" planets were discovered. Over time, however, the increasingly powerful observing machines and techniques used to resolve those problems turned up new complications in the form of visibly disorderly nonplanetary objects. What emerged in the late nineteenth century was a post-Copernican solar system full of matter in more or less orderly motion explicitly described, as I outlined in the Introduction, as relational. This view contextualized Earth as part of a cosmic environment that is neither benign nor fully separate.

From the nineteenth through the mid-twentieth centuries, the appearance of strangely behaving nonplanetary bodies disturbed the mathematical purity of an orderly cosmos, revealing the presence of boundary-crossing "matter out of place" that got marked by terms of impurity and alterity.[13] Between 1800 and 1930, European astronomers with increasingly powerful telescopes worked to correct disharmonious problems: missing planets and planetary orbital anomalies. In 1930, American amateur astronomer Clyde Tombaugh discovered the so-called Planet X, a mysterious and hidden entity that had to be perturbing Uranus's orbit; this object became the now-famous ex-planet, the dwarf planet Pluto. But previous to this, new observing regimens had been turning up other irregular objects in the solar system. In late nineteenth century, in new national observatories, comprehensive observational star charting and the introduction of long-exposure photography revealed tiny moving objects of indeterminate origin and orbit. In photographs they appeared as smears. These were the overexposed motion trails of numerous objects soon to be known colloquially as "vermin of the sky."[14] Even within scientific settings, NEOs carried the Western remnant moral load of meteors and comets, long responded to as evil auguries and called "harlot stars." NEOs' contemporary designations as "primitive bodies," "small bodies," and aster*oids* or planet*oids* (starlike or planetlike) map onto their emerging categorization as developmentally subplanetary and unsettled open systems material. These terms also suggest the need for discipline and management.

From the mid-nineteenth century to the present, the material parameters of the solar system have been constantly recalculated and recharacterized. A century after a late eighteenth-century group calling itself the Celestial Police had attempted to find the missing planet between Mars and Jupiter, nineteenth-century astronomers began cataloging the asteroids they detected as occupying a planetary-remains "belt" between the two planets. In 1950, the number of cataloged asteroids topped two thousand, with fifty-four identified as outside the main belt region. Astronomers from this point on faced the hidden rest of the solar system: a space of unknown boundaries and new classes of nonplanetary objects, both visible and theoretical. While "planets" were redefined as relatively rare, orderly, and geologically complex bodies that had largely cleared their orbital spaces of primordial hazards and debris, objects like comets and asteroids were understood as prolific, disorderly, relatively simple, and orbitally chaotic. When Pluto was reclassified as a planetoid in 2006, the International Astronomical Union also divided the broad category of nonplanetary "small bodies" into two groups: dwarf planets and a remaining grouping of asteroids, comets, and other nonplanetary objects.

As of this writing, the number of detected small bodies totals more than 740,000, and such bodies continue to be discovered and subclassified according to their locations, material morphologies, and space-traveling behaviors.[15] Astronomers have considered and dismissed theories that the asteroids of the "asteroid belt" are the remains of a broken-up planet, determining instead that all the small bodies of the solar system are captured bits of early, undifferentiated "primitive" matter, moving and interacting, not only orbiting between planets but also crossing planetary orbits. All these small bodies are being pushed and pulled by each other and by planets and set into motion through collisions, and, as a result, continue to become dislodged and create new collisions.

Indeed, the spatial boundary of a small body over the duration of the solar system's existence is, theoretically, the whole of what astronomers and my NEOphile interlocutors call the "heliosphere." To muddy heliospheric waters even more, the contemporary definition of small bodies technically includes objects of vanishing smallness, encouraging perceptions of a heliosphere crisscrossed by all kinds of material in motion headed for interaction. Boundary-crossing "small body" objects are thus not only fossil remnants of solar system formation but also evidence of a still-active impact-full solar system, raising concerns among astronomers as well as geologists, biologists, and, today, policy makers. Their boundary crossings inscribe, with their orbits and actions, the heliosphere's fully dynamic—and dangerous—material space.

## *Populations of Bodies in a More-Than-Earthly Risk Ecology*

With twentieth-century astronomical efforts to catalog, classify, and track boundary-crossing small bodies in heliospheric space came interdisciplinary perspectives that redefined planetary geological surfaces as materially both endogenous and exogenous. Astronomy and other sciences came together over efforts to understand planet/space boundary spaces like atmospheres and mysterious surface features like craters. As is typical of scientific conflicts about the interpretation of evidence across disciplinary boundaries, theories about a direct association between asteroids and planetary cratering went from controversy to mainstream acceptance. From midcentury onward, NEOs and their material properties traversed the borders of scientific and social speculation about the nature of planetary environmental boundaries.

By the mid-twentieth century, attempts to understand cosmic collision evidence on Earth and on other planets brought together disciplines that eventually constituted formal collaborative forums and spaces for what is now "planetary science." Critiques by geologists of fringe "catastrophist" theories that craters were impacts rather than volcanic remains shifted after the puzzling but devastating Tunguska Siberian impact event of 1908 called broader attention, despite lack of a crater, to the prospect that planetary atmosphere-crossing cosmic objects could be environmentally destructive. Nonendogenic geological events were not just historical but could happen now. In 1953, the journal *Meteoritics,* originally focused on the classification of meteorites, encouraged interdisciplinary opportunities to investigate nonterrestrial material and its effects, thus inaugurating a new cross-disciplinary domain (the journal later became *Meteoritics and Planetary Science*). Cross-disciplinary collaborations were steps toward breaking barriers to the production of comparative and generalizable knowledge about the Earth as a planet, signaling the appearance of what science studies scholar Peter Galison names "trading zones" for astronomy and other sciences.[16] Along with concepts like "planet," these zones include the terminology and data of "detection," "collision," and "impact."

In 1980, a watershed article in the journal *Science* sketched for a broader public the outlines of a heliospheric ecology that was constituted not only by gravitational force but also by material exchange. Physicist and Nobel laureate Luis Alvarez, in partnership with his geologist son Walter and two scientists in the Energy and Environment Division of the Lawrence Berkeley National Laboratory, combined geological, biological, and astronomical data to theorize the formation of the Cretaceous–Paleogene epochal boundary as an extinction event.[17] Using data from the Krakatoa volcanic eruption and astronomical estimations of size range among the detectible "earth-crossing asteroid" population, the colleagues concluded that an asteroid with a diameter "in the range 10 ± 4 kilometers" must have impacted Earth at the time of the event, creating, besides local catastrophic damage, atmospheric disruptions that stopped photosynthesis, disrupted food chains, and caused widespread extinctions.[18] They suggested also that other passages of extraterrestrial material through Earth's atmosphere, such as cometary ice, might have caused other extinction events. Alvarez et al.'s work seeded other attempts to investigate impacts as catalysts for earthly processes, including the "panspermia" theory of interstellar prebiotic molecular transfer that metaphorically systematizes NEOs as heterosexualized biological agents that link up with a gendered model of Earth as the Gaia mother goddess. In this reimagined interactive solar system, the Earth/space boundary becomes materially permeable.

A vulnerable Earth/space boundary became particularly relevant to human social as well as biological life during the Cold War, when the horizons of national security and capital became ever extensible through policies of military–industrial technology expansion.[19] Mitigation of the threat of cosmic impact became identified as a prospective investment site for government-funded collaboration. During the 1950s, astronomical and geological scientists interested in the behavior and impacts of small bodies began routinely sharing data and interacting with nuclear test scientists and human space program lunar reconnaissance surveyors. In 1992, brought together by charismatic authorities like "father of planetary geology" Gene Shoemaker and H-bomb "father" Edward Teller, federal agencies such as the Geological Survey and the Department of Defense collaborated in a roundtable discussion about the use of nuclear weapons to mitigate near-Earth asteroid threats.[20] This meeting signaled the concern about asteroids and comets among both scientists and the military, which led to dual-use technology proposals for military purposes and national catastrophe preparedness.[21]

By the time of the first International Academy of Astronautics Planetary Defense Conference in 2007, it had become commonplace for military planners to refer to NEO mitigation as "planetary defense," to characterize Earth's location in space as within a "shooting gallery" full of "speeding bullet[s]," and to counter biological notions that Earth's orbit is habitat sustaining by calling it "hazardous."[22] While quantification of NEO risk in such military environmental terms legitimated discussions about using space-based weaponry that might be politically opposed if the enemy were another nation, nongovernmental NEO mitigation activists soon presented an alternative narrative, arguing that appropriate anti-NEO technology cannot be weaponizable. During this period, the definition of near-Earth objects expanded to include long-period comets from the heliospheric edges and even the increasing orbital jumble of human-made objects stretching from low Earth orbit to geosynchronous satellite space. Scientific evidence that using weapons against incoming NEOs would increase their threat rather than reduce it (an exploded NEO hurtling toward Earth is not necessarily a neutralized NEO) has led to designs for nonweaponized NEO mitigation.

From the 1980s to the end of the first decade of the twenty-first century, NEO mitigation experienced a turning point as it became a topic of discussion in broader policy circles. In meetings like NASA's 1981 workshop "Collision of Asteroids and Comets with the Earth: Physical and Human Consequences," new characterizations of the relationship between Earth and space began to suggest a hierarchy of heliospheric elements ordered by their relevance to human life, and NEOs were placed along with the sun and the moon as objects of immediate consequence. The shocking discovery and close earthly pass of a "Potentially Hazardous Object" logged as 1989FC in 1989 led to two congressional study mandates: one to develop a systematic NEO detection program, the other to develop new technologies for moving or destroying asteroids. Unlike the moon, which waxed and waned as a cosmic policy object, NEOs were portrayed as an uncertain but ever-present threat to the earthly environment.

The report on a 1992 NASA workshop on NEO detection specifically stated the importance of earthly impacts to the habitable "ecosphere" as well as the importance of using technological ingenuity to avoid them, calling for a "gestalt shift" in how "humans" should think about relations with the NEO "population" and the whole solar system.[23] That study became the congressionally mandated Spaceguard Survey, in which NASA is required to characterize the material features of what the survey calls a "threat environment."[24] The survey is named for an asteroid-watch program imagined by science fiction author Arthur C. Clarke, a gesture that helped to establish what astronomer and NEO expert Stephen Ostro has described as the ongoing "blurred lines between science,

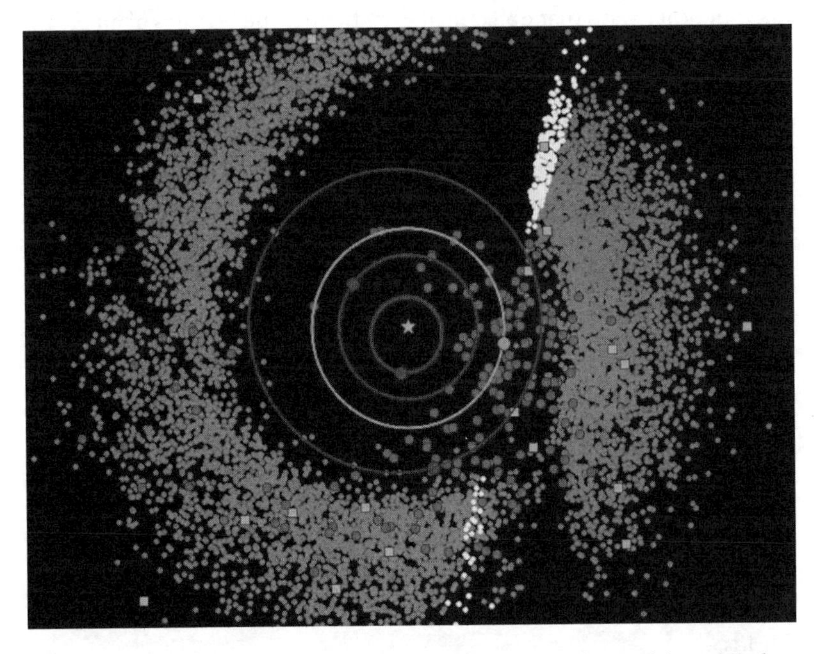

A NEOWISE mission graphic released in 2016 that renders the partial results of expert and participatory crowdsourced NEO detection as an image that resonates with modern terrestrial military radar images. Courtesy of NASA/JPL–Caltech.

futuristic thinking, and science fiction" when it comes to engagements with NEOs, which extend from detection into schemas, as I describe later, to extract wealth from them.[25] Spaceguard's charge is to detect "asteroids larger than 1–2 km," which would cause "global-scale events" resulting from the consequences of impact and structural and infrastructural collapse.[26] In this way, the survey attaches NEOs to national and earthly futures by using orbit and mass data to calculate a "quantitative estimate of the impact hazard as a function of impactor size (or energy)" and to "advocat[e] a strategy to deal with such a threat."[27]

These early 1990s activities and the dramatic 1994 impact of Jupiter by the comet Shoemaker-Levy 9 moved NEOs closer to being understood as environmental objects. That large planetary impact, the first witnessed by humans in modern astronomical history, precipitated the relabeling of NEOs as "natural hazards," which underscored the socially meaningful contiguousness of earthly and nonearthly matter. At the beginning of the twenty-first century, that meaning was elaborated in other ways that objectified even further the features of an Earth/solar system boundary made collapsible by NEO collision. In 2004, Swedish astronomer Hans Rickman and Canadian geologist and environmental catastrophe expert Peter Bobrowsky received a grant to assemble a multidisciplinary panel to address in an "open platform" the "potential psycho-social and physical consequences of a catastrophic comet or asteroid impact on Earth."[28] The resulting volume, *Comet/Asteroid Impacts and Human Society,* edited by Bobrowsky and Rickman, makes NEOs translatable as scientific policy material.

*Pace* sociologist Ulrich Beck's description of a "world risk society" in which established ideas about the spatial and temporal bounds of risk and of what counts as an enemy and as defense become unsettled, the NEO threat creates a demand for mastery of a "*worlds* risk society" in which societies are called to account for not just the figurative but also the material interconnectedness of heliospheric matter.[29] The characterization of impact risk in global terms in Bobrowsky and Rickman's volume heralds the shift of NEOs into government institutions and NGO worlds.

NEO NGO activism, in which JSC NEOphile interlocutors were involved, labels the NEO threat as more than natural. NEO activists hold that in an astronautical age, to allow a devastating environment- and evolution-altering impact or to fail to use asteroids as environment-enhancing resources is ultimately a problem of global social policy. Starting in the 1990s, popular scientific articles about asteroid and cometary threats contained cross-generational observations that the "shooting gallery" in which Earth exists makes space "not your father's solar system."[30]

At the time of my fieldwork, an animated eyeball cartoon GIF on NASA's JPL "NEO Watch" website would look around searchingly when you opened the page. That maneuvering eye represented a detecting rather than an observing gaze, illustrating the agency's commitment to monitoring the cosmos for material threats to Earth's combined environmental and social orders. This is an earthly order with broadly spherical ecological boundaries that matter to nations and international relations. In general, these boundaries include NEO paths, zones of human-made orbital debris, and incoming "space weather" (solar- and cosmic-ray) events but also extend in theory to the edges of the solar system, where life-threatening interstellar neutral hydrogen atoms push into the heliospheric bubble in varying intensities. One JPL NEO scientist has written about NEO monitoring as a continuation of the eighteenth-century Celestial Police—where policing NEOs is a moral obligation of those nations with the means to do so. As another JSC-based NEOphilic engineer told me, the appropriate selection of space environmental control technologies should be seen as an "intelligence test" for how "humans" will choose to survive and thrive in a solar system.

The development of technical systems to counter the idea of Earth-bounded systemic environmental threats—and resource limits—is at the heart of NEO spaceflight activism and its political geography and ecology of differentially exposed "human" risk and mitigation power. The FEMA simulation that the planetary defense officer told me about, involving the Department of Energy's National Laboratories, the U.S. Air Force, and the California Governor's Office of Emergency Services, revolved around a hypothetical astronomical event as a new state security scenario: a 330-ton asteroid headed toward Earth, with an expected 30-mile-wide impact destruction zone, and four years to plan.

## Defending: Securing the Political Solar System

> I think this work is most important. It feels like a calling . . . the message that comes ringing loud and clear through all the everyday nonsense and clutter is: use these unique astronautical tools to (1) explore the solar system via NEOs, (2) disperse ourselves, or (3) go extinct.
>
> —former astronaut hopeful and NEO deflection advocate, e-mail communication

NEO activism, both scientific and political, is propelled in part by privileged astronaut and cosmonaut views and viewpoints. Spacefarer views continue to play a role in global environmental and political history, including the production of "whole Earth" images and statements about a "boundaryless" world. Anthropologist Tim Ingold has pointed out how this downward- and inward-facing astronaut-on-orbit perspective of a "global sphere" authorized a Western exteriorized environmentalist sensibility that negates the spherical indwelling perspectives of others.[31]

In contrast, in NEO activism as well as in interplanetary spaceflight activism, the astronautical view displays an outward-facing privilege based on a consciousness of planet-based vulnerability—a view that is no less powerful in its shifted perspective. Outward-facing U.S. astronautical viewpoints configure a "heliospheric space" of enlightened modern perception, which aligns with the space-minded environmental perspectives of distinguished cosmological figures like Carl Sagan, Lynn Margulis, James Hansen, Stephen Hawking, and Neil deGrasse Tyson. When Phil, a NEOphile physicist and flight dynamics officer flight controller, tutored me patiently on the biological and psychological importance of perceiving terrestrial life as inside the "heliospheric bubble" of the sun's energetic influence, he sounded more like philosopher of bubbles, foam, and spheres Peter Sloterdijk than like the abstracted astronautical subject of Ingold's critique.[32] We were at a party in the northerly suburbs of Clear Lake, in the home of a NEOphile astronomer and his science fiction writer wife, where astronomers and NEOphile network associates had gathered with visiting Japanese space agency astronauts and astronomers after a local planetary science conference. During the period when I was conducting my fieldwork, NEO activism was increasing in scale and scope, becoming as vibrant as the planet-centered

U.S.-based Mars Society and the international Planetary Society, and one NEO NGO, as I describe further on, became recognized by the United Nations. This global-scale activism casts astronautical technologies as "tools" with which to manipulate an unbounded material solar system. But it also restructures the political boundaries of environmental vulnerability on Earth into NEO-vulnerable and NEO-dominating regions.

On a fall day in 2007, I sat with an astronaut/scientist/military veteran/NEO activist in a restaurant full of people working within the broad Houston-area cluster of NASA, defense contractor, and petrochemical industries. Speaking across the boundaries of his professional and spatial experience, James, a white middle-aged male, responded expressively to my standard ethnographic question "Is space a part of nature?" by making a referential shift that I found common among astronautics experts. In this shift, "nature" goes inward, becoming an innate "human" orientation to environmental control. James also evidenced what anthropologist Joseph Masco describes as the military–industrial "technoaesthetic" pleasure of enacting interventions into deep material structures and engaging in extreme boundary crossing.[33]

> You know, I'm a product of down here, so it's obviously strange when you get to space, but I never got a sense of being outside my comfort zone. I always thought in my weeks in space, I always viewed my presence there as unusual in human experience, but something that human beings were perfectly capable of enduring or even enjoying. I think it's just an expansion of the envelope that we've already inhabited and there's no, there's nothing about us that prevents us from thriving in that environment as well. In fact it's so aesthetically stimulating to be there, that it's a great pleasure to be in space, every time you look up from what you're doing you get a sense of awe or marvel, at the scenery, and just the freshness of the experience of looking out at space, black space, looking down at the beautiful Earth, it's so refreshing. . . . So, I don't see, from my point of view, a separate kind of dimension or world that we couldn't dream of inhabiting. I view it just as an expansion of our environment, and people are so adaptable, in general, in terms of going to the poles of the Earth and to the bottoms of the oceans, living on the frontier for the last few millennia, always trying to get to these

> very harsh places . . . it's very natural for me to think of humans living there for the long run. We might think of it as very exotic today, but I don't think it's gonna be that way in fifty years.

Earth and human beings, in this view, are not separate from outer space, just environmentally "down here."

During this interview, I asked James to clarify a phrase I heard him use in reference to the NEO–human relationship in a presentation at NASA Headquarters in Washington, D.C., several months earlier. The meeting was part of the NEOphiles' ongoing effort to promote their "humans to NEOs" mission concept through pitches to publics and decision makers; I had been following this effort in the press and through direct observation. In his talk, James summoned a powerful image of future astronauts "manipulating with their own hands and machines something that could destroy us." In response to my request for clarification, he explained himself by quoting an Apollo-era astronaut who initiated a petition within a spaceflight NGO, the Association of Space Explorers, for an international NEO detection and mitigation policy:

> Well, Rusty Schweickart has said it many times, and I agree with him, we're alive at the time when we actually have just developed the technology to prevent us from being snuffed by an asteroid, and we've been subject to extinction for as long as humanity's been around, with absolutely no ability to do anything about it, and for the first time in history, our species is actually able to turn off a process that has been running the solar system since it began. So it's actually a pretty historic moment, to actually think of altering the way the solar system operates, in a way that preserves our ability to survive. So, our technology has actually caught up with the dynamics of this process and we're on the verge of being able to manipulate our way into some survival, in a way that the dinosaurs didn't have the capacity to do. So rather than being subject to the whims of celestial dynamics, we're now able to make sure that we don't get waxed by this process, [the] bombardment that's been going on. And I think it'll be very satisfying to actually have people nudging these bodies around or operating around them and really demonstrate that we have the capacity. It'll be quite momentous.

Here NEOs are portrayed as agents of perverse violence within a rationally reengineerable solar system.

This story conveys a politically systemic view of Earth, in which "we" and "us" emphasizes a whole species facing a NEO together, not a collection of unequally prepared societies facing their own NEO-mitigation threats. If, as another NASA NEO expert, Donald Yeomans, notes in a *New York Times* op-ed piece, "We owe our very existence and current position atop the food chain to these celestial visitors," then the capacity to access and manipulate them reveals a corresponding sociopolitical "chain" of national capacities for ecosystemic security and expansion.[34] The comprehensive ecological vision of a species avoiding the natural cataclysmic cycles of its biological niche maps master narratives of adaptation onto progressive narratives about global leadership in environmental planning.

Both James and his astronaut colleague Rusty Schweickart belong to the Association of Space Explorers, an NGO embedded in governmental and spaceflight networks aiming to change the way the Earth–NEO relationship "works" in order to influence the coproduction of social and natural orders.[35] They warn in universalizing terms that human access to astronautical technologies makes NEOs a fundamentally social rather than natural problem, and that a refusal of nations to unite to mitigate this problem amounts to a global social and, ultimately, human evolutionary failure. The ASE is a nongovernmental professional and spaceflight advocacy organization made up of "over 400 flown astronauts and cosmonauts from 37 nations."[36] Such distinctions have earned the ASE and its elected leadership observer status in the United Nations Committee on the Peaceful Uses of Outer Space. Following a rhetorical pattern common in extreme environment preparation discourse, the ASE takes the characterization of the Earth–NEO relationship from an *as if* notion (as if an asteroid could hit the Earth) to a *what if* scenario (what if one is headed to do so). In 2008, the ASE's Near Earth Objects Committee drafted a formal petition to the United Nations. While the ASE petition reflects what social scientists Stephen Collier and Andrew Lakoff describe as "a profusion of plans, schemas, techniques, and organizational initiatives that respond to new kinds of perceived threats to collective security," it is also an activist document that declares both action and nonaction to be political and represents astronautics as a developmental process by giving human evolution temporal scalarity from the Paleolithic to an unknown future and from a global surface to unknown heliospheric ends.[37]

The B612 Foundation, another NGO concerned with NEO impacts, was cofounded by Schweickart and is named for the asteroidal home of the title character in Antoine de Saint-Exupéry's novella *The Little Prince.* In keeping with Saint-Exupéry's depiction of a prince wise beyond his youthful mortal appearance, the foundation's name invokes the value of thinking about NEOs transcendentally across social process boundaries. In 2007, during my fieldwork, the B612 Foundation website invited visitors to "join" in the "ultimate environmental project," which the foundation members started because of their dissatisfaction with the "current lack of action" by the U.S. government to protect the Earth from near-Earth asteroids.[38] The foundation's goal was to "significantly alter the orbit of an asteroid in a controlled manner by 2015," using technologies such as a "gravity tractor" to push or pull a near-Earth object into a less threatening orbit. As chair of the foundation, Schweickart represented its goals at public and governmental forums by expertly characterizing what the website called "the current environment." The website described this environment as not the static "old solar system" but a dynamic space in which the Earth and its "sister planet," the moon, inhabit a "neighborhood" heavily populated with small bodies that cross their orbits.[39] In 2007, Schweickart gave testimony before a congressional subcommittee, as he has done frequently, reminding the audience that "NEOs are part of nature" and calling into question the utility of thinking about NEO impacts as accidents that individual nations should plan for. In doing so, he advocated for global, preimpact prevention in the name of "evolutionary responsibility."[40]

In the ASE petition to the United Nations, Schweickart and his associates frame the goal of "planetary defense" as a universal moral imperative to develop technocratic schemes of environmental protection planning based on international preagreements about mechanisms of response, mitigation, and, ultimately, regional sacrifice. As this petition was being drafted, the media reported on ongoing controversies over NEO threat calculations and policies, including a controversy concerning the asteroid 99942 Apophis. The asteroid, spotted in 2004, was named, a NEOphile told me, in honor of a character in the science fiction television series *Stargate SG-1,* an alien who uses a portal to cross spatial boundaries in order to destroy and enslave. The character is named for the Egyptian light-eating god of utter destruction. While its orbit was being confirmed, 99942 Apophis was graded as a "potential threat" in 2036, if it passed through a kind of mathematical stargate called a "keyhole" in 2029. NASA scientists reported that if the NEO, now classed more specifically as a PHO (potentially hazardous object), passed through this orbital time/space plot, it would collide with Earth with a force at a TNT equivalent of 880 megatons. Boosted by depictions of asteroids in popular culture as destroyers as well as global uniters, discussion of Apophis grew, and the asteroid quickly became a boundary object of thought and practice. Its threat became potentialized in military science, meteorology, geology, and public policy as well as numerology ("It is expected to pass Earth the first time on April 13, 2029—a Friday the 13th. What's more, 2 + 0 + 2 + 9 = 13").[41] In 2009, Apophis was downgraded to a nonexistent threat for its pass on 2029 and "ruled out" as a threat in 2036, but its behavior when it passes Earth on both dates, as close as some geosynchronous satellites, remains unpredictable.

NEO activists use the threat of total destruction to converge the boundaries of scientific and moral prediction, calling for advance decisions about how to protect spaces from impact. The ethics of taking responsibility for spatially tracking NEOs becomes a politics of drawing spaces at low or high risk of impact. In 2008, the ASE petition to the United Nations became the basis for a report titled *Asteroid Threats: A Call for Global Response,* which sets out legal and policy justifications for such responses by reinterpreting some key terms of the 1967 Outer Space Treaty.[42] The report links the treaty's reference to the importance of scientific "planetary protection" from potentially harmful extraterrestrial material with an associated treaty resolution in favor of using and sharing "remote sensing" information to prevent "a threat to 'the Earth's natural environment.'"[43] Acting responsibly, however, requires two things: a plan to try to deflect the NEO and the international capacity to adhere to prearranged political and even military agreements between potentially affected regions. Such agreements are deemed necessary because deflection technologies, whether nonweaponized nudges by spacecraft or nuclear detonations, may end up simply changing the course or place of impact.

The ASE petition acknowledges that nations are unequally equipped to participate in deflection enterprises and goes so far as to suggest, in highly abstracted terms, that the Earth has to be assessed in terms of protection-deserving spaces and those that must "absorb" risk and damage.[44] Deservingness, long attached to social markers, becomes attached to spatial ones. For this "rational" scheme to work, each participating nation must bear an individual responsibility to recognize and prenegotiate its own actions and survival, based on invocations of "international law" and liability procedures that prevent states from withholding deflection plans or "knowingly" using deflection technologies to damage other states' "territory."[45] The document is laced with references to the "humanistic" responsibilities of an "international community," but the multisystemic outlines of a NEO deflection politics based on unequal technology possession, the power of beneficence or hostility, and the emergence of a new liability culture become visible in the narrative.

In the context of this highly charged moral ecology of systemic risk, NEO activist elites continue to engage in what they and my NEO-phile interlocutors call "NEO awareness" campaigns extending into the global South. In 2014, the B612 Foundation became involved in the Asteroid Day global awareness campaign started by Queen guitarist-turned-physicist Brian May and his artistic and scientific associates. The Asteroid Day organization is underwritten by the signatures of "hundreds of esteemed scientists, physicists, astronauts, and Nobel Laureates from 30 countries and leaders in business and the arts" on a petition to accelerate NEO detection and deflection projects and to support public and educational programs with "regional coordinators" in sites alphabetized from Afghanistan to Venezuela. A video series available on the Asteroid Day website, narrated by U.S. physicist and public intellectual Neil deGrasse Tyson, is divided into segments on NEO "detection," "characterization," and "international" dimensions, which emphasizes the production of broader global information-sharing and trust relationships.[46]

The Asteroid Day campaign adds to an elite moral ecological mapping of unequal environmental protection and risk statuses, which now extend upward from surface depths and surfaces, through national airspaces, and far beyond the atmospheric domains of "climate." Included in this schema are astronautical technologies to remake the ideological and material scalarity of "human" environmental risk and security. NEOs go from being objects of chaos to being objects of objective concern, as things made public through elite cultural consensus about processes of morally motivated calculative and political characterization and as things that, as historian of scientific objectivity Lorraine Daston argues, appear universal and free of idiosyncrasy.[47]

As technical environmentalist icons, NEOs are charismatic megamatter. Like the atmospheric carbon load and warming seas, NEOs have an ecosystemic life constituted by numbers, temporal and spatial threat calculations, authoritative spokespersons, and activisms—including the human mission activism I followed at JSC—that consolidate the social dimensions of their meaning.

## The New Solar Ecosystem: Human–NEO Relations

Some NEOphiles advocating for a human exploration mission to a NEO during the Constellation program began their institutional and public presentations with a PowerPoint slide containing this proviso:

> NEO Study Disclaimer:
> This is only a Phase 1 technical feasibility study.
> NASA has not endorsed this mission concept yet.

This slide communicated the conceptual nascence and organizational liminality of their mission concept. Their Phase 1 (early mission design proposal) study, titled "A Piloted Orion Flight to a Near-Earth Object: A Feasibility Study," was based on the Orion crew capsule's intended multisited "moon, Mars, beyond" capabilities.[48]

The feasibility being tested was both technical and political. Through continued development by its authors and supporters, the study began to gain the features of a "concept of operations"—a military and systems engineering phrase that refers to a "high-level" (general) description of an operational project's "notional" (ideal) architecture and systems integration. One manager in the Engineering Directorate, a middle-aged white woman who spent a career writing procedure manuals, told me that a "con-ops" could not be just a technical plan but had to "tell a story" about a mission's total value. NEOphile activism values NEOs as multipurpose material that conjoins systems with spatial and temporal scales: biological, technical, economic. While this narrative calls for fixing NEOs into structures, it also portrays them as open systems objects of emergent potential. It does so by championing a human rather than robotic mission, signaling an understanding of spaceflight potentiality as manifest in direct embodied human–extraterrestrial engagement rather than in the distant detection and defense fixing of Earth-based NEO relations.

The JSC NEOphiles were scattered—some in the Astromaterials Resource and Exploration Science Division building, some in the Astronaut Office, some in life sciences, some in the Mission Operations Directorate working on flight controller consoles. Some frequented the Explorogroup Thursday-morning meetings in the Astromaterials Division building, where "small body" astronomer Richard, one of the Phase 1 study authors, had an office. On the door of that office was a photograph of Stephen Hawking taken when he visited NASA after riding in the "vomit comet" aircraft microgravity simulator. Richard shared his office with people working on lunar science, interplanetary dust collection, and orbital debris tracking. He explained to me that within the social hierarchy of astronomers, cosmologists are on top and planetary astronomers are at the bottom, meaning that NEO astronomers occupy an even less exalted spot. While his description validates Hans Blumenberg's historical diagnosis that post-Copernican astronomical thought maintains the professional ascendancy of philosophy and theology in the form of cosmology, it also points to how astronomers can be divided in terms of their devotion: to observation only or to human planetary exploration.[49] Richard and the NEOphiles often groused that their astronomical and astronautical colleagues did not value NEOs for either observation or exploration because they were "illiterate" about NEOs. Most people at NASA, they explained, had no idea what near-Earth objects or asteroids were, or they assumed that they were objects in the asteroid belt closest to Earth. But Richard and his colleagues exploited the mysterious and awe-inspiring materiality of NEOs in their presentations.

In their slide presentations, NEOphiles framed NEOs as accessible and manageable "stepping-stones" for a species that will eventually need more resources than the Earth offers. This kind of material-limits claim is common to environmentalists and space exploration and exploitation activists. The "Phase 1" category of the NEOphiles' study gave its authors license to speak speculatively and with a kind of conceptual freedom about their hopes for NEOs as sources of provisioning materials. The mission campaign used images and language that elaborated an understanding about the basic nature of the NEO–human relationship that surpassed the disciplinary bounds of astronomy and geology, heading into the economic and metaphysical. With NEOs as things that materialize an unbounded space of environmental security and exploitation, progress in human space exploration gets mapped onto human evolutionary progress in the extreme.

In spring 2007, the NEOphiles were perfecting their presentation activism. I went to watch several of them give a PowerPoint talk on the Phase 1 study to a local JSC and Houston-area audience at the Lunar and Planetary Institute. LPI is housed in a small, low complex of brick buildings northwest of Johnson Space Center. Opened in 1968 as a planetary science research center that also provided support for JSC's lunar operations, LPI was originally operated jointly by Rice University and the National Academy of Sciences. Its management was then taken over by the NAS's Universities Space Research Association, a consortium of one hundred member universities with space science and engineering programs. Since fall 2006, "impact theory of life" originator David Kring had been managing LPI's new web-based lunar science information portal, and other scientists at the institute were engaged in projects focused on nonplanetary bodies and material, from comets and Kuiper belt objects to interstellar dust. The version of the presentation the NEOphiles were giving in 2007 was one they would go on to give elsewhere, including at a major science fiction convention, where NEOs were seen as having fantastical potentials as sites of inhabitation and alternative utopian bases. All NEOphile presentations drew from a repository of slides shared among the feasibility study authors and their managerial stakeholders. The many NASA presentations of various kinds I saw over the years were government property subject to export control and other kinds of restrictions, but they were also used to cross institutional and public boundaries to gain support for ideas.

Joking about the popular conception of asteroids as alien threats, the presenter, NEOphile astronomer Richard, averred, "We have no plans to wage war on asteroids." He then anchored the mission concept with a bibliographic slide that legitimated it as an old spaceflight idea. He and the Phase 1 study authors created bigger historical bounds for their idea, linking it to an Apollo-era study to mount a Saturn V rocket mission to the asteroid 433 Eros in 1975 and ending with two recent studies done during the Space Exploration Initiative in the early 1990s, including a paper by astronaut and study team member Tom Jones. Richard then specified the constraints under which the study's authors labored: to prove that the mission was possible without major modifications to Constellation's current launch and crew capsule systems architecture.

Whether the audience was astronomically literate or "illiterate," Richard and other NEOphiles prefaced their descriptions of the mission by explaining NEO ontology as transgeological and NEOs as made up of "primitive" materials vital to new understandings of solar system formation and to future Earth system resourcing. This had a professional boundary-work purpose as well. They were campaigning for public recognition of a new model of heliospheric dynamics and materials, and they knew they were introducing many people to the idea that there are other cosmic objects that could orbitally interact with Earth. They were also introducing the idea of an open-ended human ecology. As Star and Griesemer describe, boundary objects have to "reconcile" differences in meanings and worldviews, and in this case, what had to be reconciled was not only what kinds of objects NEOs are but also how they reveal materially permeable solar systemic boundaries.[50] As a result, the study presentations became ways for NEOphiles to communicate with NASA employees and publics about the contemporary solar system and to situate NEOs within the cultural conceptual boundary of significant cosmic things.

To put NEOs into context, NEOphile presentations used illustrations that quantified and classified NEOs as near, prolific, and without predictable boundaries. In his spring 2007 talk at LPI, Richard showed slides depicting how NEOs vary in size, including a photographic image of the Japan Aerospace Exploration Agency's Hayabusa mission target, the asteroid Itokawa, next to the equivalently sized Golden Gate Bridge. NEOs, he showed, can be round, can made up of rubble, and can be covered with smaller rocks that are the size of meteors that have made visible holes on Earth (one slide showed Arizona's Meteor Crater). Richard's next slides illustrated how NEOs are detected and classed as threats, but these were followed by slides that touted NEOs as destinations for humans in terms of science, understanding of the solar system "neighborhood," and sources of "volatiles" (oxygen and hydrogen) that could provide life support and fuel for spacecraft. The slides that set up a context for this had a dramatic impact, as, with a series of clicks, Richard modified the normal image of the inner solar system environment. To a simple black-and-white graphic image of a sun orbited by four planets, successive slides added more and more colored dots representing an increasing "NEO population," until the previously blank spaces were full of color representing matter: a "crowded" solar system. This drove home, with cognitive and affective impact, the notion of an Earth surrounded by visible matter, not invisible space, and installed in the audience's minds the notion of astronautically accessible asteroids as open-ended resource providers rather than government program resource users.

In these ways, the study and the NEOphiles' presentations argued that NEO defense, exploitation, and exploration are naturally interrelated. With enthusiastic claims about the dual "practicality" of NEO detection and exploration, Richard asked the audience to rethink the cognitive and emotional givens that make Earth and space appear separate

A "mission concept" image of a human NEO mission from the website of the JPL Center for Near Earth Object Studies. Courtesy of NASA/JPL–Caltech.

and environmentally separate, and to consider how a "human presence" on a NEO would be a significant step toward resourcing the heliosphere for science, markets, and movement away from environmental limitations. In conversation with NEOphiles and other astronautics practitioners, I was not surprised to find that they had read or were reading geographer Jared Diamond's books on the relationships among social formations, technologies, and environmental mastery or collapse and cited them as evidence for their cases. In their minds, NEOs represent objects to access in the name of avoiding collapse.

As environmentally distributed objects more spread out and easy to access than planets, as the NEOphiles emphasized, NEOs were often categorized as alternative spaceflight destinations, along with "Lagrange" or "libration" points. These are not objects but mathematical points of gravitational balance that occur in the dynamics created by three objects in space, also called the "three-body problem." If one of those objects is a spacecraft in motion and another is a planet or the sun, these points can be used to remap the solar system topologically for spacecraft trajectory plotting, turning a problem into an equation in momentum transfer. NASA mathematicians and mission designers propose future mission trajectories using an "interplanetary superhighway" that they can plot by taking advantage of libration point gravitational dynamics that create "low-energy" pathways for efficient spacecraft fuel use and travel. JPL mathematician Martin Lo describes this highway as "a vast infrastructure provided by the Solar System just like the Jet Streams in the atmosphere or the Great Currents on Earth. As all natural resources, once discovered, it must be developed, mined, and harvested."[51] In this vision, the gravitational fields of planets and other moving bodies, including NEOs, become energetic catalysts, providing gravitational "halos" that create ever-shifting subway-like "tubes" that can propel vehicles from one energy site to the next. The NEOphiles plot trajectories to NEOs that do not follow this superhighway. But they do ask audiences to rethink the solar system resource landscape, where asteroids become closer and more useful than they appear.

Richard and his fellow NEOphiles capped their presentation at LPI with a series of bullet points showing the importance of NEOs' inherent interconnections with "exploration," "science," and the "general public"— categories that reflected language NASA leaders were using to "sell" the

Constellation program to publics. The NEOphile presenters highlighted the value of the human explorer qua human and reminded the audience that a human–NEO relationship already exists and that it calls for deliberate rather than accidental knowledge and prospective management. Beginning with "exploration," Richard invoked the outward-facing astronautical perspective ("We will be leaving Earth behind") as well as the opportunity for the efficiency testing of deep-space destination hardware. For the "general public" rationale, Richard reminded the audience of what it would mean socially to leave the Earth/moon system, to explore inner solar system space, to make progress toward Mars, and to characterize a potentially hazardous object up close.

NEOphiles took versions of this "paper NASA" PowerPoint story on the road in formal and informal ways. They told me how they would get into NEO conversations on planes, in restaurants, at academic gatherings, and in any setting they could. Their "mission" was to (as some NEOphiles put it) "convert" people. Such "conversion" often involved passionate statements; for example, a NEOphile mission control technician told me one day in the JSC cafeteria that "single-planet species, intelligent or otherwise, do not survive." NEOphiles shared with news media science writers their individual stories about the genesis of the NEO mission. Among the publications including such stories that came out while I was doing my fieldwork were Gregg Easterbrook's article "The Sky Is Falling" in the June 2008 issue of the *Atlantic,* about governmental complacency toward the NEO threat (this piece was later included in *The Best American Science Writing, 2009*); a cover story on the crewed mission option in the July 2008 issue of the Smithsonian magazine *Air & Space*; and *Star Trek* science consultant André Bormanis's article for the November/December 2007 issue of the Planetary Society's *Planetary Report* titled "Worlds Beyond," a piece as much about the "asteroid underground's" struggle to get the concept of NEOs recognized as it is about the concept itself.[52]

This public storytelling constitutes a fervent space in which people aim to move NEOs from their systemic peripheries to the center of American security and space policy debates, but it also traces the imaginative contours of a fully human-inhabited solar system. As a kind of "alternate universe," Timothy's "NEOverse" contains a critique of NASA planet-centric programs, legitimating activists' centering of NEOs in social and biological progress. Their work is about engaging not only in what Bruno Latour calls lengthening a "network" but also in what sociologist Steven Epstein calls a "credibility struggle."[53] In such a struggle, people gain support for scientific claims by having knowledge that appears to be empirically credible not just within the domain of science but also in the face of controversy about what kinds of science can and should be done—in this case, controversy about the future of U.S. human spaceflight as a form of U.S. big science and of humans as a scientifically aware but ecologically imperiled cosmic species. The NEOphiles I spoke with were quick to joke about how "the dinosaurs didn't have a space program" and to assign NEOs to "stepping-stone" status for spaces as well as for processes. In the ecological mode of describing seemingly harmful environmental things as beneficial in other perspectives, NEOphiles saw NEOs as other than doomsday material.

## Resourcing: NEOs, New Space, and Infrastructure in the Solar Ecosystem

> Read Heinlein's book, *The Moon Is a Harsh Mistress.* [In it] people are shipping raw materials back to Earth, and end up dropping raw materials on strategic targets on the Earth. I like Heinlein, Haldeman, etc., they write hard sci-fi. . . . The point is that lunar raw materials might be useful if you can ship them to a factory. I'm just not seeing what a "colony" can provide that will be useful to the home planet.
>
> —Ken, space life sciences administrator

> Late last night coming home from [the airport], a guy [on the airport shuttle] with me asked what I did at NASA . . . of course the NEO thing came up and this business dude was hooked. Hooked.
>
> —NEOphile, e-mail communication

In mid-twentieth-century American and Soviet science fiction stories, asteroids are imagined to be sites of enclosed colonies or rough and tough mining or energy extraction projects. The latter trope is visible in the 1998 movie *Armageddon,* which features an anti-asteroid mission training scene in JSC's huge spaceflight training pool, the Neutral Buoyancy Laboratory. In the movie, roguish oil-rig roughnecks are selected as having the right stuff for a mission to an incoming asteroid; the plan is for them to drill explosives into the asteroid so that it will explode before it can hit Earth. This is one imaginative chapter in what NASA sociologist Linda Billings describes as an "American national narrative" that unifies "frontier pioneering, continual progress, manifest destiny, free enterprise, and rugged individualism."[54] Even if they are not threats, asteroids can be ontologically reduced to raw dirty rocks or to raw materials. As nonplaces, asteroids orbit outside the planetary legal bounds of territorial sovereignty and even environmental protection. To the eye of industrial capital, asteroids' unearthly and unclaimed rawness places them on a potential pathway to an industrial but not necessarily colonially settled solar system. During the Obama administration, NEOs became markers on a policy pathway to connect science, security, and national industrial resourcing. Continuing the Bush administration's plan to create what NASA administrator Mike Griffin called a "Solar System Economy," Obama-era NEOs were boundary material for a solar ecosystem, in which the historical conceptual relationship of economy and ecology was reinstated. As economic resources, NEOs also bridge the conceptual boundary of "system" and "infrastructure," becoming fixed into the role of structure-building materials. As the quote from Ken above suggests, if space sites cannot be colonized by human bodies, they can be incorporated through an exploratory infrastructure focused on U.S. national desires to occupy free space and on capitalism's ideal material: free stuff.

If some NASA leaders thought the NEOphile human-to-NEO mission story was a "waste of time" to tell, NEOs became multipurpose con-ops protagonists under Obama-appointed NASA administrator Charlie Bolden, a veteran astronaut. After the Obama administration's declaration that the Bush-era "Program of Record" (formerly known as Constellation) was in crisis because of "goals that do not match allocated resources," in 2009 the Review of U.S. Human Spaceflight Plans Committee recommended what it called the "Flexible Path" to nonplanetary sites without "deep gravity wells"—such as NEOs, the moons of Mars, or Lagrange points—to "extend our presence in free space."[55]

> The Flexible Path represents a different type of exploration strategy. We would learn how to live and work in space, to visit small bodies,

> and to work with robotic probes on the planetary surface. It would provide the public and other stakeholders with a series of interesting "firsts" to keep them engaged and supportive. Most important, because the path is flexible, it would allow for many different options as exploration progresses, including a return to the Moon's surface or a continuation directly to the surface of Mars.[56]

This is a scaled-up version of the cultural concept of "flexibility" that anthropologist Emily Martin finds in the U.S. social valorization of a fit human immune system.[57] It also reflects the astronautical theorization of society as an environment for spaceflight systems, which have to be adaptable to public demands for novelty and the exploration of new space.

As exploratory objects with speculative value, asteroids have also become material for contemporary ideas about how to respond to planetary and economic limits. Soon after Alvarez et al.'s article on the K–Pg boundary extinction appeared, NASA's 1983 Infrared Astronomical Satellite mission began to return systematic spectral data on 1,811 asteroids, confirming ways to classify them according to composition. Using such data, University of Arizona planetary geologist John S. Lewis speculated about the trade value of asteroid materials. His claims followed on the asteroid wealth dreams fueled in the 1960s by futurist NASA engineer Dandridge Cole's dramatic assertion that it would be possible to extract "$50,000,000,000,000 from the asteroids."[58] In the 1990s, Lewis became a spokesperson for commercial asteroid mining, arguing that evidence of pure undifferentiated materials promised "untold riches" that could justify the expense of space exploration.[59] As with enthusiasms for harnessing cosmic energy from lunar helium-3 or orbital solar collectors, the proponents of asteroid mining treat cosmic resources as forms of economic and exploration-sustaining providence.

In 2012, pro–NEO mission astronaut Tom Jones joined a cadre of entrepreneurs, academics, and scientific and military experts in launching an asteroid mining start-up called Planetary Resources. The move to privatize the materials mined from asteroids gained legal formalization during the Obama administration, which mandated support of "New Space" entrepreneurs creating commercial spaceflight investment projects. My fieldwork in space conferences brought me into proximity with these entrepreneurs, who were often vocally derisive of NASA, including its failure to exploit space resources. The U.S. Commercial Space Launch Competitiveness Act, signed into law in 2015, sketched out a "road map" past the Flexible Path, requiring presidential support for solar system economic development, including private ownership of extracted solar system resources—leaving intact the international prohibition of territorial rights. Under Administrator Bolden, NASA also mobilized teams to move through study phases for the robotic Asteroid Redirect Mission, with the goal of extracting a boulder from a NEO and bringing it into Earth orbit for sampling and study. Planetary Resources was among a set of companies preparing for water and metals extraction. Such exo-geoprospecting, like bioprospecting, proposes to remake the marketable scope of nature and the chemical scope of markets, but it is also aimed at unbounding the terrestrial systematicity and imaginative scalarity of industrial capitalism.[60] In business plan stories, space-based asteroid mining is the "next step" for industrial evolution that protects earthly ecological resources while opening up the pursuit of unlimited economic material.

As with JPL scientists' plans for a solar system transportation infrastructure, rhetoric about using asteroids as building materials is where I encountered the most consistent use of the word *infrastructure* within otherwise "system"-saturated astronautical discourses. Calling asteroidal water the "oil of our solar system" and speculating about how asteroid sources of rare earth metals could transition societies to new forms of energy, NEOphiles often described NEOs as the raw materials for a fully functioning "spaceflight infrastructure." These advocates recognize, as anthropologist Timothy Mitchell argues in his examination of a political economy dependent on oil, the historical relationships among material forms, industrial infrastructures, and kinds of economies.[61] But NEOphiles add to this the capacity of NEOs to restructure the bounds of human inhabitation—what NEOphile physician Kyle called "building infrastructure for a spacefaring civilization." In chaining together ideas about asteroids as providential ecological objects and infrastructural "building blocks," NEOphiles echo what Daniel Worster, in his history of the relationship between economic and ecological thought, describes as eighteenth-century "political and economic as well as Christian" theories of nature as a "chained-together" machine subject to divine and human management.[62] NEOs today appear as far-ranging solar systemic parts with an eschatological role in conjoining human systems into structures. Meanwhile, underneath New Space entrepreneurial speculation lies the tangible world of spaceflight materiality on Earth, in which business plans and utopian dreams chain together a profitable solar ecosystem.

## Space Matters

"Tektite," the physicist said, picking up a charred-looking shard of meteoritic glass and handing it to me in his Astromaterials Division office, "you find them in East Texas." While I was doing fieldwork, NEO astronomers spoke about recovery missions to find bolides that had landed in deserts, and JSC astronauts in Antarctica found burned dark meteorites sunk into the blue-white ice. Counting its natural and socially-produced materials, outer space has a rich materiality on Earth. During my fieldwork, those materials circulated from hand to hand on earthly ground: meteorites, papers, mission patches and stickers, pieces of the moon,

NEOWISE "edge-on view of our solar system" rematerializes the solar system as a cloud of matter rather than as planets arranged on elliptical planes. Courtesy of NASA/JPL-Caltech.

T-shirts, gear for breathing underwater, reports, space food, parts of spacecraft, funeral memorial programs, equipment for shipping to and from research sites, pieces of habitat mock-ups, spacesuit helmets, vials of astronaut saliva, procedure manuals. Early in my fieldwork, people in the Astromaterials Division were working on the Stardust Mission to comet Wild 2. Chunks of aerogel sent into space were, for the first time, capturing cosmic dust for return to Earth. The mission also rematerialized astronomer Carl Sagan's famous invocation "We are made of star stuff." After the craft crash-landed back on Earth in 2006, photographs of its undamaged gel surfaces entered the Stardust@home citizen science project, in which people connected themselves via their home computers into a crowdsourced systematic search for the microscopic impacts of interstellar dust grains. Spaceflight's materials assemble a social cosmos.

Proliferating forms of robotic and human spaceflight make, for social groups with access to them, a postterrestrial field of open systems and potential infrastructures. This field makes new scopes and scales of sociality perceivable—in their unlimited potentials and contained limits, and in their hard bounds and aspirational unboundedness. As systems material, NEOs bridge Earth and space geology in a way that renders a three-dimensional political geography of space access and vulnerability. They travel across boundaries as a reminder of how space things become Earth things and vice versa—and how space programs are Earth programs.

Among the many space programmatic materials I was handed during my fieldwork were autographed astronaut photos. I asked astronaut candidate Yvonne Cagle for one when I met her in her university lab in fall 2006. These photos index practices of stardom and ceremonial gifting. But they also serve as messages from spaceflight-ready hands. Cagle is an African American woman with a long career as a spaceflight expert and advocate, and when I met her, she was working on technologies to bridge spaceflight, Earth biomedicine, and national security. Opening the sunlight-dappled trunk of her car in a parking lot strewn with pine needles and redwood pollen, Cagle pulled her official NASA portrait out of a folder. She found a Sharpie, signed her name, and wrote out her personal signature phrase: "Space for all."

Then and now, I read Cagle's words as links between the domains of statement and question, testimony and invitation. They open a space of inquiry into the processes that continue to socially relate but also separate things as systemic parts of a space within which life and non-life emerge and transform together. Cagle's words invite the question of how to imagine the future of Earth and its forms of "all" within this systemic space but also to be ever mindful of *who* and *what* systemic visions are being made for.

# CONCLUSION
## FUTURE SPACE

In this book I have focused on how contemporary U.S. government-funded spaceflight programs expand the kinds and boundaries of "environments" while they control and govern "systems" within them. I have treated "system" as an ethnographic object—a *relational technology*—to examine socially interconnected, but not always ideologically or ethically aligned, efforts to naturalize as well as build sets of relations. Human spaceflight is a process that produces technical systems but also extends metasystemic processes—cultural processes that make all systems look to be analogous, to be governed by the same kinds of relational rules and developmental processes, and to be coconstitutive of human social and ecological subjectivity. In spaceflight, connecting bodies, technologies, and sites like planets and asteroids as systemic parts enlarges a total human environment made up of norms and extremes. In this environment, human habitation opens up as a transitional and technically improvable condition. All of these processes contribute to constituting the national political systematicity of material and symbolic connections and separations, as well as empowering people to imagine and enact systemic change.

The purpose of this ethnography is to bring "environmental systems" into a new kind of attention. In the book's chapters, environmental systems are spatial formations that exist in the intersections of seemingly contrary conditions like givenness and constructedness, homeostasis and emergence, enclosure and openness. Spaceflight practices evince how systems of living and nonliving things are designated as well as made by social groups in ways that, in turn, enable perceptions of relational interdependence as well as discreteness. Systems appear to be already real and yet they are also technologies of reality. Environments appear to be intimately connected to bodies but also to extend into disconnected and unperceivable spaces. This is, therefore, an ethnography of contradictoriness as much as it is an anthropological study of spaceflight's environmental systems.

For scientists and engineers as well as administrators and political leaders today, "system" is less useful when it is deployed only as a relational technology of enclosure; it is more powerful when it is worked with as a provisionally open-ended and aspirational form. The contingently semiclosed/semiopen system can be used to enforce understandings of environmental boundaries and to unbound them. My ethnographic fieldwork on these environmental relation-building processes spanned Democratic and Republican presidential administrations that were seemingly opposed in their commitments to "the environment" as a shared endangered space. But within government and industrial contract sites where people are employed to work in *and* through U.S. political regime changes, "environment" and associated concepts like "ecosystem" are increasingly technocratized: they are spaces of systems to be managed and capitalized. As a result, U.S. spaceflight is about more than outer space and flight. Its systems instantiate a relational scalarity that crosses ideological and political party boundaries. These days, in part because of spaceflight, to call any space—natural, social, political, economic—an "environment" conveys meaningful notions of a more or less open space filled up with systems of reworkable relations.

While I did fieldwork, all kinds of events had impacts on NASA's systems and environments: the proliferation of orbital craft and debris, extreme weather events, mission successes and catastrophes, the cancellation of the space shuttle and Constellation programs, and presidential administration changes. The engineers, scientists, administrators, and astronauts with whom I interacted in different sites were anxious about and critical of administration changes and what they meant for the future of robotic and human spaceflight as a long-term process. Thousands of people lost their jobs at NASA during these administration changes, but others made careers out of "surviving" them (to use one lunar scientist's word). People I spoke with were watching NASA's programs grow and shrink, the institution age and lose relevance, and its industrial partnerships wax and wane. They noted the new entrepreneurial energy of commercial spaceflight, with its more youthful and socially varied workforce and its leaders' libertarian rather than civil attitudes toward outer space's potential.

Despite presidential leadership changes, the United States has not retreated from outer space as a site of strategic national forms of spatial control, from scientific exploration to military operations. U.S. spaceflight historians Roger Launius and Howard McCurdy hold that there is a "myth" that presidential leadership determines spaceflight development. They argue that presidential leadership went from an "imperial" command role in the 1960s to its contemporary supporting role in relation to what they describe as the flourishing of military and commercial space "infrastructures" and "systems" that are independent of executive visions.[1] Their argument echoes what people working in NASA told me during the first year of Donald Trump's presidential administration: what is at stake during administration changes is not just particular human or robotic programs but also, as one engineer put it, "systems in place" as well as "systems to build." The ongoing processes of political systematicity and environmental expansionism that I have described in this book continued to be visible not only during the unveiling of Trump administration space policy but also in the environmental imaginaries of American science fiction movies that spanned the Bush, Obama, and Trump spaceflight eras.

Even as the early Trump administration positioned itself against the Obama-era "environment" as a site of industrial threat and began to dismantle environmental regulations, it supported continued spaceflight system building in the context of what analysts called emerging national "space systems security" projects to "protect" U.S. military and commercial interests in the "space environment." During such presidential transition phases, veteran and newly minted experts on spaceflight share their diagnoses of the political scene on space-focused news sites and blogs. In May 2017, European space policy analyst Maximilian Betmann posted a piece on emerging U.S. space strategy on The Space Review, a popular spaceflight news analysis blog that many of my interlocutors read to find what one engineer described to me as "well-grounded" commentaries.

In his post, Betmann describes how U.S. military and civil authorities have been intently responding to new "counterspace" threats (antisatellite technologies, including mechanical interference and signal interference) to its "space systems."[2] This began following passage of the National Defense Authorization Act of 2014, which specifies "an identification of offensive and defensive space control systems, policies, and technical possibilities of future systems," including "an identification of any gaps or risks in existing space control system architecture" as well as new modes of "space situational awareness."[3] Space situational awareness beyond Earth requires increasingly sensitive and specific sensor systems attuned to changes in the material and energetic conditions of space, such as spacecraft positions and space weather (magnetic field and radiation changes that affect technical systems). The technical extension of human bodily senses into outer space makes it an everyday part of elite environmental perception. Betmann concludes, "Another important aspect of new U.S. strategy is the idea of making its space systems more resilient so that they can continue to operate in a contested environment."[4]

Both the National Defense Authorization Act and Betmann's analysis display metasystemic analogical and processual modes of systems work, in which both the mechanical and the organismic parts of systems can be made more developmentally resilient and predictive. Space's transition to being a "contested environment" for the United States in social and natural terms involves security issues, but it also includes the increase in orbital debris that extends Earth's ecosystemic pollution and resources. As articulations of this continuing "threatened" and "contested" space environment were appearing on space news sites and blogs in 2017, NASA scientists alerted me to announcements about meetings being scheduled to respond to Trump administration policies, including the fifty-fifth Robert H. Goddard Memorial Symposium, titled "Future Space: Trends, Technologies and Missions," which would feature talks focused on a "one NASA" approach to uniting science (basic and natural science) and exploration (human spaceflight programs), despite the historical shifts in funding focus that usually come with changes in presidential leadership and (as one session title called it) the "political environment."[5] Upon signing the first NASA authorization bill of his presidency, which continued the agency's Obama-era funding levels, President Trump stated that the bill "reaffirm[ed] our national commitment to the core mission of NASA," of human spaceflight and technoscience, but also that it was "about jobs"—highlighting the existence of opposing opinions about the agency as an elite jobs program and aerospace industry tax-funding source. However, in contrast to reading this as a signal that NASA would inaugurate a new interplanetary human spaceflight program, Scott Pace, a space policy analyst, called the signing a "vote for stability."[6] Such stability keeps jobs and systems in place.

Trump administration space policies feed the ongoing systemic politics of Earth–space connections and separations—not as planet and cosmos but as strategically conjoined environments. During Trump's 2016 campaign, his advisers promised that if elected he would curtail NASA's "politically correct environmental monitoring" and restore the agency to its outward space-facing mandate. Despite news reports that Trump had admitted to believing there may be "some connectivity" between human action and climate change, media rumors abounded that the administration would turn climate satellites around in order to generate more space-facing science knowledge.[7] In spring 2017 the Trump administration declared its intention to end NASA's "Earth-centric" science, thereby delimiting the systemic machinery of climate change science. However, Trump's nominee for NASA administrator, Oklahoma congressman Jim Bridenstine, stated that he saw value in studying climate change on Mars, since studying "other planets can inform our understanding of Earth."[8] Hidden in that statement is a political systemics perspective I heard from a handful of NASA scientists wanting more-than-terrestrial knowledge about terrestrial climate change—that dramatic planetary climate change, including Earth's, could be attributed to larger solar or galactic systemic changes. All of these statements highlight not just the contemporary politics of Earth–space environmental connections and separations but their history as well, such as the pivotal role NASA's comparative Venus–Earth–Mars atmospheric science missions played in the development of "greenhouse gas" and climate change science.

Contemporary acts of contingent Earth–space connection and separation create new political ecological Earths. The Earth of Trump's Republican-supported policies is emerging as a planet with a less comprehensibly knowable interior atmospheric environment, but with a potentially more knowable exterior environment. U.S. orbital space is slated to become less civilly scientific and more militarized and privatized, producing new obfuscations of public knowledge about planetary environmental stability and change. However, the political systematicity of Earth–space environmental relations also results from Democratic administration policies to restrict outer space science and exploration programs and focus on, as one Clinton-era NASA program explained it, "missions to planet Earth." NASA's ongoing role as an environmental agency as well as a space agency also connects it in more technically contingent and politically fraught systemic ways with other environmentally focused agencies, such as the Environmental Protection Agency and the National Oceanic and Atmospheric Administration. These processes highlight the increasingly extreme political systematicity of a post-terrestrial human environment—in actual as well as fictional U.S. science and technology.

The joy of attending new science fiction movies and then passionately debating their relative technical and environmental accuracy is a staple of NASA work life, but what motivates these activities are people's passions for questions about the transitive habitability of Earth and space. These questions are haunted by technoscientific predictions about an increasingly uninhabitable Earth and a potentially uninhabitable outer space. My fieldwork included long talks with groups of people about movies like *WALL-E* (2008), an eco-critical animated film about an abandoned robot tasked with industrial cleanup while a spacecraft full of systemically coddled humans awaits the possibility of return to a decontaminated Earth, and *Gravity* (2013) and *Interstellar* (2014), movies with very different plotlines but shared displays of spaceflight systems building within a hostile outer space. The television series *The Expanse,* which began airing in 2015 on the Syfy cable network, was popular because it gathers together decades of American science fiction portrayals of a racially, socioeconomically, and militarily divided solar system, including the ecobiopolitics of Earth's relations with social groups living off-world, whose bodies, access to potable water, and physical debasement as workers make them ecosystemically politicized and vulnerable. In all of these depictions, space is the environmental background for Earth as the proper but jeopardized human home. Commenting on such movies and television shows, a NASA life scientist and an astronomer both reported to me how they hated "wrong" depictions of physics and technology but also said that they appreciated any movie or TV series that points out what one called, ambiguously, the "awfulness of real outer space." I also remember my fascinated shock when an elderly Apollo-era astronaut informed me with experienced conviction that, in fact, "the moon sucks" as a destination for future human habitation. That humanity's proper home is the Earth or the solar system is not a neutral fact that simply indexes biological belonging; it also invokes the discourses and material processes by which social groups empower or dispute that fact—and how they subsequently produce authoritative knowledge about the dimensions of real, actual, and future human space.

During my fieldwork and since that time, the planetary environment that continues to be a target for human transhabitation planning is Mars; these plans are infused with both ecological and economic aspirations for a future inhabitable solar ecosystem. As I have pointed out in this book, the Trump administration and its circles include people directly or indirectly associated with space entrepreneurship, such as Steve Bannon and Elon Musk, who perceive capitalism as a transenvironmental process of elite controlled material transformation and distribution. While entrepreneurs like Robert Bigelow, the aspiring space hotelier, have joined other space capitalists in funding orbital space interests, Musk continues to advance his designs on Mars. Musk's electric car and new energy development factories and Mars-forward tunnel-building efforts to reduce traffic congestion in Los Angeles colocate with the other major producer of Martian futurism in the United States: Hollywood. During 2015, as NASA workforces struggled to recover from the Obama-era cancellation of Constellation and underfunded research into what it would take to retrieve an asteroid for scientific study, upper-level administrators seized opportunities to associate the agency's readiness for interplanetary travel with the release of the movie *The Martian.* Based on the best-selling book by Andy Weir, the movie is, according to space life scientists and engineers, unusually technically accurate. It also features a futuristic vision of a well-funded NASA, with state-of-the-art facilities and commercial partnerships focused on making Red Planet missions happen. When the movie came out, it had a governmental promotional life through NASA, as relations between the agency and the filmmakers resulted in aligned images of speculative efforts toward permanent human inhabitation of Mars.

These visions illustrate the systemic imbrication of U.S. spaceflight projects as acts that connect actual and speculative environments. While in this book I have described the ways in which such work occurs in state-sponsored systemic annexations of outer space, I have also followed the active aspirations of people working transversally to that work—keeping what counts as collective space open to question and action. The solar system's political ecology is both material and hypothetical, and it is produced through actions and dreams of systemic conjoining and bounding. These acts take multiple forms, from presidential decrees and agency programs to smaller-scale actions by people producing outer space visions and discourses. In one such systemic vision, a free neoliberal man stands at the center of an economic cosmos without boundaries, distanced from a host of less free people and searching for more and more valuable stuff. In another vision, a plural but more justly integral human collectivity works toward a postterrestrial future space for each and for all. At the heart of such visions is a visionary scalarity—a micro to macro sensibility of social futures as more than earthly.

How people work with "system" as a relational technology enacts sensibilities about how and in which ways things can or should be related. This book you hold in your hand or behold on a screen is a product of information and industrial systems. But it can also be understood in terms of a scientific story about how you and it came to be together and apart. How far from the sun or Pluto is this book or are you, in pixels and atoms, if you think of all of these things as interconnected rematerializations of solar and galactic power and materials? Human spaceflight, whether it is labeled as something right or wrong to do, extends fleshly bodies into that space of inquiry, and into the uneasy question of where environmental relations begin and end.

# ACKNOWLEDGMENTS

I AM GRATEFUL TO MANY PEOPLE; this work lives in the world through their goodwill and generosity. Spaceflight practitioners welcomed me into their workplaces and social worlds with enthusiasm and patience. I hope all will feel my intention to portray their work respectfully. I am beholden to the wise, perseverant guidance of John Charles, Mark Craig, Edna Fiedler, Rob Landis, Wendell Mendell, and the late Judy Robinson, who made the fieldwork possible. I thank the National Space Biomedical Research Institute and my indefatigably supportive supervisors, Edna Fiedler and Pam Baskin, who brought me on board the NEEMO space analog missions managed by NASA and National Undersea Research Center personnel, all of whom were willing to let me participate in depth.

My appreciation continues for the many others I met in space life and physical sciences, flight medicine, human factors, architecture, habitability design, engineering, the astronaut corps, history offices, and communications departments in government and private sectors. I will always remember with gratitude all the people who opened doors for me at NASA Headquarters, Johnson Space Center, the Ames Research Center, and the Jet Propulsion Laboratory, as well as the good-humored spacefarers who kept in touch from beyond Earth's surface. I experienced kindness and unflagging toleration from the inimitable members of the JSC Explorogroup network, who said yes to interview requests and brought me into organizational inner worlds. The NEOphiles taught me, on land and at sea, unforgettable lessons about space-impacted life on an ever-transforming Earth. Steve Dick, Steve Garber, Roger Launius, John Logsdon, and diligent historians and archivists in the NASA Headquarters history office graciously supplied documents and experience-based insights that helped me to contextualize my fieldwork. Scholars in the JSC oral history project and at the University of Houston–Clear Lake, including Jennifer Ross-Nazzal and Mark Scroggins, identified crucial textual materials. Contractors and participating members of NASA's industrial and professional association network, including those in the American Astronautical Society, the American Institute of Aeronautics and Astronautics, SAIC, the Space Frontier Foundation, and Wyle Laboratories, were often vitally informative companions in work sites and at conferences. I thank Alan Thirkettle, the European Space Agency's ESTEC Noordwijk Technical Center in the Netherlands, and the European Astronaut Center in Köln, Germany, for accepting my internships there.

All of these spaceflight sites provided me with comparative perspectives on astronautical history and practice, enabled as cosmic and earthbound spacefarers brought me through gates, doors, and airlocks to watch them work with resolute commitment. *Ad astra per aspera.*

From the beginning, James Faubion at Rice University steered me steadily into the project's conceptual foundations; that this book exists at all is because he, Eugenia Georges, Chris Kelty, Hannah Landecker, George Marcus, Cyrus Mody, Amy Ninetto, and Julie Taylor believed in its feasibility and value. The project was also formed in conversation with Dominic Boyer, Cymene Howe, Jeffrey Kripal, Patricia Reiff, and Stephen Tyler. I shared the student's journey with Michael Adair-Kriz, Casper Bendixsen, Jae Chung, Trevor Durbin, Hank Hancock, Terri Laws, Elise McCarthy, Nahal Naficy, Michael Powell, Brian Riedel, and Dan White. I found my first environmental space with David Seibert, Pete Little, and Rodrigo Rentería-Valencia. The project's fieldwork and writing were supported by a Lodieska Stockbridge Vaughan Fellowship, a Wagoner Foreign Studies Scholarship, a Rice Space Institute grant, a National Science Foundation Dissertation Improvement Grant (no. 06020442), and a Rice Humanities Research Center Dissertation Writing Fellowship.

Outer space anthropology is a small world, peopled by deeply congenial colleagues with whom small and big ideas generatively converge and diverge. Debbora Battaglia, Stefan Helmreich, Lisa Messeri, Michael Robinson, David Valentine, and Deanna Weibel were generous advisers and allies, helping me build and rebuild the project's ethnographic and theoretical scaffolding as they engaged their own outer spaces. As they worked in related areas, Joe Masco, Patrick McCray, David Nye, Steven Pyne, Cynthia Selin, Sharon Traweek, and Matt Wolf-Meyer provided pivotal counsel on engaging with government science and technology at its embodied, spatial, and temporal edges.

The manuscript came into its final form through encounters with scholars in a variety of venues, including the American Anthropological Association Body & Time panel (2004); the Rice University Humanities Research Center Mellon Seminar (2007–8); the Bringing STS into Environmental History Workshop in Trondheim, Norway, organized by Sara Pritchard, Dolly Jörgensen, and Finn-Arne Jörgensen (2010); and the Oil & Water working group, with Andrea Ballestero, David Hughes, and Nikhil Anand, sponsored by the Center for Ethnography at the University of California, Irvine (2014).

My extraordinarily convivial colleagues at UCI provided me with vibrant intellectual camaraderie and profoundly influenced this project's maturation and completion. I am joyfully indebted to Victoria Bernal, Tom Boellstorff, Geoff Bowker, Mike Burton, Dave Feldman, Kim Fortun, Mike Fortun, Angela Jenks, Eleana Kim, Bill Maurer, Michael Montoya, Keith Murphy, Sylvia Nam, Simon Penny, Damien Sojoyner, and Mei Zhan. Students closely pursuing problems of system, ecology, environment, extremity, and cosmology were wonderful to think with: Emily Brooks, James Goebel, Ségolène Guinard, Anirban Gupta-Nigam, Beth Reddy, Simone Popperl, and Peter Taber.

The transformation of manuscript into book was a collective process for which I will be ever thankful. Jason Weidemann and the board and staff of the University of Minnesota Press produced this over-the-top project with warm and skillful human-scaled stewardship. Diligent reviewers gave me invaluable perspectives, and I benefited greatly from their generative pushback and rich suggestions. Glenn Swanson, space studies scholar and former Johnson Space Center historian, subjected the final manuscript to a keen historical and factual review. Amy Ninetto and David Valentine, mentors to me from the beginning, fine-tuned the work with disciplinary knowledge and wise writerly expertise. Judy Selhorst copyedited the final draft with accomplished grace and care.

Family and friends tolerated my being in outer space for far too long, and I could not have made it without their steady support. During the trip, foundational friendships were sources of joyful calm refuge. Michael Adair-Kriz, Andrea Ballestero, Andrea Frolic, Kris Peterson, Roberta Raffaetá, Marek Reavis, and Guy Zimmerman were lifelines. My sisters Ann Barbarow, Regan Overholt, and Cynthia Steineke had my back. John Boone accompanied and helped me through the process. Phyllis Heft nudged me on with tenacious love and belief. My parents, Gail Olson and Bruce Olson, started all of this by tucking me into a green canvas butterfly chair when I was five to watch the grainy live Apollo 11 broadcast from the lunar surface in 1969. They encourage my curiosity about different spaces then and now.

This book is for my daughter, Morrigan Olson Bruce, who is my light and inspiration.

# NOTES

## Introduction

1. Historians and philosophers are tracking this phenomenon (see the citations in the notes that follow). The list here refers to Talcott Parsons's systematic application of systems theory in structural-functionalist social action theory; sociologist Niklas Luhmann's foundational theory of interlocking cognitive and material social systems, with its epistemological developmental framing of ongoing "second-order" levels of observation and understanding; Clifford Geertz's anthropological rendering of culture and religion as systems and ecosystems as containing humans; and Hans-Jörg Rheinberger's influential depiction of experimentation as a system that relies as much on human technique as on the agential emergence of new epistemic things that in turn shape further experimental goals and experience.

2. Haraway and Harvey, "Nature, Politics, and Possibilities," 512; Gibson-Graham, *The End of Capitalism.*

3. See Law, *Power, Action, and Belief*; Latour, *Aramis*; Starosielski, *The Undersea Network*; Star, "The Ethnography of Infrastructure"; Von Schnitzler, "Traveling Technologies"; Jensen, "Experimenting with Political Materials"; Larkin, "The Politics and Poetics of Infrastructure"; Ong and Collier, *Global Assemblages*; Marcus and Saka, "Assemblage."

4. By focusing on how system became a conceptual and material technology of relation making, I follow social theorists, historians, and social scientists who are concerned with varieties of *techne,* such as the blends of social and

material technologies that legitimated scientific experimentation, the "technologies of the self" involved in making a social identity, the ways in which living substances like cells and genes are made to work as biotechnologies, and how the "virtual" is both a space and a tool of social experience. See Shapin and Schaffer, *Leviathan and the Air-Pump*; Foucault, *Technologies of the Self*; Landecker, *Culturing Life*; Franklin, *Dolly Mixtures*; Boellstorff, *Coming of Age in Second Life*; Kelty, *Two Bits*; Galloway and Thacker, *The Exploit*; Carrier and West, *Virtualism, Governance and Practice.*

5. See Rasch and Wolfe, "The Politics of Systems and Environments," parts I and II. These issues of *Cultural Critique* provide a foundation for critical systems studies in articles and debates by a set of scholars who work reflexively with second-order cybernetics theory to delineate the boundaries and potentials of posthumanist theory.

6. See Olson and Messeri, "Beyond the Anthropocene."

7. I derived these statistics at the end of the Constellation program in 2010 by working with "data cubes" (no longer available) on the website of the NASA Shared Services Center, http://nasapeople.nasa.gov. As of this writing, these data are accessible at NASA, "Workforce Information Cubes for NASA," accessed December 15, 2017, https://wicn.nssc.nasa.gov.

8. Stengers, "The Cosmopolitical Proposal."

9. I use *map* to indicate that my ethnographic aim is to create a way to perceive systemic extensivity as much as it is to describe it, *pace* Geertz, "Thick Description."

10. Gupta, *Red Tape,* 32.

11. Mills, *The Power Elite,* 223; Das, "The Signature of the State," 225.

12. W. Henry Lambright notes the emergence of NASA as an environmental agency in his essay "NASA and the Environment," which is based on his earlier work, *NASA and the Environment.*

13. The cyborg study has been reprinted, along with an excellent analysis of its history and meaning by Chris Hables Gray. See Clynes and Kline, "Cyborgs and Space."

14. See N. Katherine Hayles's historical analysis of U.S. posthumanist engineering and informatics cultures, in which she argues that cybernetic models of futuristic humanness animate Western liberal dreams of a disembodied transcendental self. Hayles, *How We Became Posthuman.*

15. NASA Technical Reports Server, search conducted June 5, 2017, https://ntrs.nasa.gov.

16. Tresch, "Cosmogram."

17. System has a "hidden" history as a "tradition" in Arnold, "Systems Theory." It is a conceptual historical "puzzle" in Faber, "Whitehead at Infinite Speed."

18. Levine, *Forms.*

19. Siskin, *System.*

20. Strathern, *After Nature,* 76.

21. Strathern once named "the relation" as a metaphorical figure that allows connections and comparisons to be made from a position within a world understood to be on the same "earthly plane." See Strathern, *The Relation,* 18. In later work she references the production of astronomical interworldly relations, citing Shapin, *A Social History of Truth*; and Biagioli, *Galileo's Instruments of Credit.* See Strathern, "Reading Relations Backwards."

22. Strathern, *The Gender of the Gift*; Strathern, "Reading Relations Backwards," 14.

23. Jones, "System Symptoms," 116.

24. Foucault, "Of Other Spaces"; Rancière, *Disagreement*; Deleuze, *The Fold*; Latour, "Give Me a Laboratory"; Rheinberger, *Toward a History of Epistemic Things.*

25. Strathern, "Reading Relations Backwards," 14.

26. Blumenberg, *The Genesis of the Copernican World,* 540–64.

27. Hoyningen-Huene, *Systematicity,* 26–27; Franks, *All or Nothing,* 10.

28. See Moeller, *Luhmann Explained*; Palumbo-Liu, Robbins, and Tanoukhi, *Immanuel Wallerstein.*

29. See Helmer, *Schleiermacher and Whitehead*; Stenner, "James and Whitehead."

30. See Webb, "The Planet Mars and Science"; Lane, "Mapping the Mars Canal Mania."

31. Doel, *Solar System Astronomy in America,* 231–32.

32. The most socially influential work on systems at the microscopic level can be found in Maturana and Varela, *Autopoiesis and Cognition.* For a discussion of the transdisciplinary influence of Maturana and Varela's work, see Mugerauer, "Maturana and Varela." The leading proponents of the macroscopic social scientific applications of system are Talcott Parsons, Margaret Mead and Gregory Bateson, Niklas Luhmann, and Immanuel Wallerstein; for an example of emerging world and Earth systems synthesizers, see Hornborg and Crumley, *The World System and the Earth System.*

33. See Hans-Georg Moeller's examination of Jürgen Habermas's critique of Luhmann's systems thinking as animated by the idea of life as "for itself." Moeller, *Luhmann Explained,* 22.

34. Among the analyses implicitly positioned against systemic forms and thought, the following are relatively explicit: Haraway, *Simians, Cyborgs, and Women*; Merchant, *The Death of Nature*; Gibson-Graham, *The End of Capitalism*; Harvey, *The Condition of Postmodernity*; Biersack and Greenberg,

*Reimagining Political Ecology*; Shohat, "Notes on the 'Post-colonial'"; Appiah, "Is the Post- in Postmodernism the Post- in Postcolonial?"

35. Meltzer, *Systems We Have Loved.* For a collection of critical assessments, see Shanken, *Systems.*

36. Faber, "Whitehead at Infinite Speed"; Deleuze and Guattari, *Anti-Oedipus*; Deleuze and Guattari, *A Thousand Plateaus.*

37. Johnson, *The Secret of Apollo.*

38. Arnold, "Systems Theory," 12.

39. See Checkland, *Systems Thinking, Systems Practice*; Meadows, *Thinking in Systems*; Hammond, *The Science of Synthesis*; Ackoff, *Redesigning the Future*; Churchman, *The Systems Approach and Its Enemies.* On Bell Labs, see Gertner, *The Idea Factory.*

40. See Hagen, *An Entangled Bank*; Taylor, "Technocratic Optimism." On assembly rules, see Golley, *A History of the Ecosystem Concept,* 28. Golley's book is a masterful work of conceptual and social history.

41. See Hayles, *How We Became Posthuman*; Turner, *From Counterculture to Cyberculture*; Farman, "Speculative Matter"; Haraway, "A Manifesto for Cyborgs."

42. Jasanoff, "Ordering Knowledge, Ordering Society," 14.

43. Logsdon, *Legislative Origins of the National Aeronautics and Space Act,* 30, 32–33.

44. Carson, *Silent Spring,* 6–8.

45. Arendt, *Between Past and Future,* 273.

46. Ibid., 260.

47. Golley, *A History of the Ecosystem Concept,* 69–74.

48. Organisation for Economic Co-operation and Development, *The Space Economy at a Glance 2014.*

49. Budgetary figures come from the NASA fact sheet *NASA FY 2017 Budget Request.* During the Apollo program, NASA received approximately 4 percent of the national budget. Many of my interlocutors cited this percentage as a benchmark for how much civil astronautics was seen to matter.

50. See Handberg, *Reinventing NASA*; Launius and McCurdy, *Robots in Space*; Lambright, "The Political Construction of Space Satellite Technology"; Lambright and Schaefer, "The Political Context of Technology Transfer"; Mindell, *Digital Apollo.* For "big picture" treatments, see Burrows, *This New Ocean*; McDougall, *The Heavens and the Earth.*

51. Dick and Launius, *Societal Impact of Spaceflight.*

52. See, for example, Benjamin, *Rocket Dreams*; Atwill, *Fire and Power*; Tribbe, *No Requiem for the Space Age*; Lavery, *Late for the Sky*; Klerkx, *Lost in Space*; Dean, *Leaving Orbit.* Among the many program histories are Cooper,

*Before Lift-Off*; Smith, *The Space Telescope*; Shirley, *Managing Martians*; Burrough, *Dragonfly.*

53. Appel, "Offshore Work"; Easterling, *Extrastatecraft,* 15.

54. Cosgrove, *Apollo's Eye*; McDougall, *The Heavens and the Earth.*

55. Braun, "Environmental Issues," 644.

56. Limerick, "The Adventures of the Frontier"; Redfield, *Space in the Tropics,* 49; Redfield, "The Half-Life of Empire in Outer Space."

57. Messeri, *Placing Outer Space.*

58. Ruzic, *Spinoff 1976,* 35.

59. Kroll-Smith, Couch, and Marshall, "Sociology, Extreme Environments and Social Change," 3, 4.

60. For work on planetary sphericality, see Ingold, "Globes and Spheres"; Jasanoff and Martello, *Earthly Politics.* See also DeLoughrey, "Satellite Planetarity and the Ends of the Earth."

61. Luke, "On Environmentality"; Launius, "Writing the History of Space's Extreme Environment."

62. Said, "Performance as an Extreme Occasion," 331.

63. See Lovelock, *Gaia*; Ward, *The Medea Hypothesis*; Lovelock, *The Revenge of Gaia*; McKibben, *Eaarth.*

64. Jan Zalasiewicz, quoted in Carrington, "The Anthropocene Epoch."

65. Helmreich, "Extraterrestrial Relativism."

66. See Rabinow, *French Modern*; Janković, *Confronting the Climate.* As I describe in chapter 2, historian of biology and medicine Georges Canguilhem argues that the medical category of the pathological derives its meaningful power not as the opposite of health but as a perturbation that stimulates organismal self-regulation. Canguilhem, *The Normal and the Pathological.* Janković shows how the Victorian-era concerns of the English with the role of environmental control in managing social vitality and behavior are at the root of today's environmental medicine and policies to regulate interior spaces and bodies.

67. Robinson, *The Coldest Crucible.*

68. Tuan, *Topophilia.*

69. Bainbridge, *The Spaceflight Revolution.*

70. Nicole Stott, in "Exclusive: Astronaut Nicole Stott—Interview: Part 1," YouTube, November 7, 2014, https://www.youtube.com.

71. Battaglia, *E.T. Culture.*

72. See Olson, "The Ecobiopolitics of Space Biomedicine"; Wolf-Meyer and Taussig, "Extremities."

73. Marburger, keynote address.

74. Worster, *Nature's Economy.*

75. Golley, *A History of the Ecosystem Concept,* 189, 196.

76. Nisbet, *Ecologies, Environments, and Energy Systems,* 6. Nisbet's history of ecological and land art provides a window into the ways in which artistic interests in closed and open systems produced various modernist representations of built and natural environmental systematicity.

77. See Jasanoff and Martello, *Earthly Politics.* In particular, within this edited volume, see Lahsen, "Transnational Locals."

78. Escobar, "After Nature."

79. Clark, *Inhuman Nature,* 29.

80. Olson and Messeri, "Beyond the Anthropocene"; Swyngedouw, "Depoliticized Environments."

81. See Lambright, *NASA and the Environment*; Lambright, "NASA and the Environment"; Hersch, *Inventing the American Astronaut.*

82. Valentine, "Atmosphere."

83. Anker, "The Ecological Colonization of Space."

84. See Battaglia, Valentine, and Olson, "Relational Space." On environmentalisms, see Little, "Environments and Environmentalisms in Anthropological Research."

85. Bebbington and Bury, *Subterranean Struggles.*

86. Biersack and Greenberg, *Reimagining Political Ecology.*

87. Tsing, *The Mushroom at the End of the World.*

88. Howitt, "Scale as Relation."

89. Siskin, *System,* 36–42.

90. Clark, "Derangements of Scale."

91. On the relational imagination, see Morten Pedersen in Venkatesan et al., "The Task of Anthropology Is to Invent Relations," 61.

92. I mean to underscore the term *actual* in the sense described by Paul Rabinow, as an orienting framework for an anthropology of the contemporary and its ethnographic pursuit of objects of study as apparatuses, problematization, conjunctures, and events. See Rabinow, *Anthropos Today.* I oppose the idea that the "real" astronautics is military; instead, I hold both civil and military together as deliberately delineated parts of government and industrial environment-making apparatuses.

93. I derived these 2010 statistics by working with "data cubes" (no longer available) on the website of the NASA Shared Services Center, http://nasapeople.nasa.gov. As of this writing, these data are accessible at NASA, "Workforce Information Cubes for NASA," accessed December 15, 2017, https://wicn.nssc.nasa.gov.

94. Comprehensive demographic information is not available for contractor pools as it is for civil servants—this is one result of the political strategies of governmental systematicity.

95. It may be possible to speculate on some interlocutors' identities based on events/settings (such as public gatherings), so I have made efforts to disguise such indicators. I may have gone overboard in taking such precautions, and some individuals may have preferred to be identified directly, but it is my sincere intention to be sensitive to the conditions of the interlocutors' generous participation.

## 1. Metasystem

1. The term *analog* is also rendered in NASA gray literature as *analogue.* I use the former spelling because it is the most commonly used in technical documents.

2. NASA, *NASA's Exploration Systems Architecture Study,* 59, 222.

3. For a significant representation of the metasystem concept in systems engineering, see Hall, *Metasystems Methodology.*

4. Ortner, *Sherpas through Their Rituals,* 152, 162.

5. Urban, *Metaculture,* 32.

6. Lee, foreword, xi.

7. See Boyer and Marcus, "Introduction," 6.

8. Messeri, *Placing Outer Space,* 67–68.

9. NASA, *Human Research Program,* 51.

10. The use of systems as forms of expert knowledge and the sets of things within which women and black subjects were included and from which they were excluded can be found throughout two seminal works on this topic: Paul and Moss, *We Could Not Fail*; and Foster, *Integrating Women into the Astronaut Corps.*

11. Baudrillard, *Simulacra and Simulation.*

12. For a historical contextualization of this announcement, see Lambright, *Why Mars.*

13. The difficulty of tracing the history of system practice is a concern expressed in systems histories as well as in practitioners' attempts to locate the origins of their discipline. An important contribution to this history can be found in Brill, "Systems Engineering."

14. Chapman, *Project Management in NASA.*

15. Hughes and Hughes, "Introduction," 22.

16. Pickering, "Systems Engineering at the Jet Propulsion Laboratory," 125, 129.

17. Ramo, "The Systems Approach."

18. Miles, "Introduction," 2.

19. See Melosi and Pratt, *Energy Metropolis.*

20. See Limerick, "The Adventures of the Frontier"; White and Limerick, *The Frontier in American Culture.*

21. John F. Kennedy, "Address at Rice University on the Nation's Space Effort," September 12, 1962. https://www.jfklibrary.org.

22. Logsdon, *Legislative Origins of the National Aeronautics and Space Act,* 50.

23. Trachtenberg, *The Incorporation of America,* 26, 33.

24. See Stepan, "Race and Gender."

25. See Holyoak and Thagard, *Mental Leaps*; Hallyn, *Metaphor and Analogy in the Sciences.*

26. Addelson, "The Man of Professional Wisdom," 22.

27. Weddle, *Argument,* 7.

28. Nye, *America as Second Creation,* 13.

29. Sténuit, *The Deepest Days,* 19–21.

30. National Space Biomedical Research Institute, "About" and "News," accessed April 1, 2010, http://www.nsbri.org.

31. Intuitive Surgical, "Company: History," accessed October 6, 2016, http://www.intuitivesurgical.com.

32. Anthropologist Matthew Wolf-Meyer has traced how sleep monitoring fits among the assemblage of processual "parts" involved in the modern systematization of economies and bodily processes. See Wolf-Meyer, *The Slumbering Masses,* 19.

33. NASA, *Human Research Program Integrated Research Plan,* 73, 111.

34. Diprose, "Biopolitical Technologies of Prevention."

35. See Vertesi, *Seeing Like a Rover*; Bass, Wales, and Shalin, "Choosing Mars-Time"; Shirley, *Managing Martians.*

36. For a view into the ways the concept of "operations" scopes and scales military action beyond battlefield sites, see Masco, *The Theater of Operations.*

37. Munn, "The Cultural Anthropology of Time," 116; Helmreich, *Alien Ocean.*

38. Fabian, *Time and the Other,* 177n19. Fabian calls attention to Hans-Georg Gadamer's note that the history of the system concept is related to "attempts to reconcile the old and the new in theology" and "the separation of science from philosophy." Ibid., citing Gadamer, *Wahrheit und Methode.*

39. Ibid., 120–21.

40. Grosz, *The Nick of Time,* 51–52.

41. Farr, Christensen, and Keith, "The Business Case for Spiral Development," 2, 12.

42. Strathern, *Partial Connections*; Haraway, "Situated Knowledges."

43. Bateson, *Steps to an Ecology of Mind,* 267.

44. See Jenkins, *Confronting the Challenges of Participatory Culture,* 3–4.

45. I found archived NEEMO 3 mission reports on NASA's NEEMO site (https://www.nasa.gov/mission_pages/NEEMO), but early NEEMO mission reports are no longer available. This report was titled "Journal 3—Danny Olivas: Mission Day 0: Saturday, July 13, 2002."

46. Ibid.

47. Siskin, *System,* 76.

48. Paul and Moss, *We Could Not Fail.*

49. Shetterly, *Hidden Figures.*

50. Chandler, *X—the Problem of the Negro,* 60.

51. Morgan Watson, quoted in Paul and Moss, *We Could Not Fail,* 196.

52. Paul and Moss argue that the Huntsville, Alabama, spaceflight center was an exceptionally successful site for black professionals, despite Governor George Wallace's efforts to intervene. They quote from a Wallace speech in which he insisted that the future of "our government system" was imperiled by desegregation, and they also discuss the extent to which the system concept was an element of the discourse about national integration and international competition. Ibid., 140.

53. See Luke, "On Environmentality"; Agrawal, *Environmentality.*

54. National Research Council. *A Framework for K–12 Science Education.*

55. On invitational rhetoric, see Foss and Griffin, "Beyond Persuasion."

56. Kamler, *Surviving the Extremes.* The volume's Penguin edition is subtitled *What Happens to the Body and Mind at the Limits of Human Endurance.*

57. Kamler, "To Practice for Mars."

58. DNews, "Finding NEEMO."

59. See Haraway, "Cyborgs and Symbionts"; Garb, "The Use and Abuse of the Whole Earth Image"; McGuirk, "A. R. Ammons and the Whole Earth"; Jasanoff and Martello, *Earthly Politics.*

60. See Odum, *Ecological Vignettes,* 108.

61. Ibid., xiii.

62. Poynter, *The Human Experiment.*

63. Robinson, *Future Primitive.*

64. Choy, *Ecologies of Comparison.*

65. White, *The Overview Effect.*

66. Suedfeld, "Space Memoirs."

67. In her masterful essay "Objectivity and the Escape from Perspective," Lorraine Daston discusses the origin of scientific aperspectival objectivity in moral philosophy, a process that is resutured within "overview effect" advocacy (see chapter 5).

68. Mae Jemison, in *Planetary.*

69. Beer, *Darwin's Plots.*

70. Phillips, *The Bends,* 2.

71. "Journal 13—Mike Gernhardt: Mission Day 7: Saturday, October 27, 2001." As with Danny Olivas's mission journal cited above, I accessed Gernhardt's journal at the NASA NEEMO website, which no longer makes early mission reports available.

## 2. Connection

1. Traweek, *Beamtimes and Lifetimes.*

2. Latour, *Science in Action.*

3. Dyson, "Introduction," 5.

4. Hersch, *Inventing the American Astronaut.*

5. NASA, "Man-Systems Integration Standards," sec. 3.2.1.

6. NASA, "NASA Space Flight Human-System Standard," sec. 1.4.

7. Ibid.

8. See Gerhardt, *The Social Thought of Talcott Parsons.*

9. Wynter and McKittrick, "Unparalleled Catastrophe for Our Species?," 30.

10. Rabinow and Rose, "Biopower Today," 215.

11. Ibid., 204, 215.

12. Rose, *The Politics of Life Itself*; Paxson, "Post-Pasteurian Cultures"; Helmreich, *Alien Ocean.*

13. Griffin, "The Space Economy," 2.

14. Some material in this chapter first appeared in Olson, "The Ecobiopolitics of Space Biomedicine."

15. Cooper, *Before Lift-Off.* This carefully documented book is full of accounts of astronauts professionalized in relation to systems mastery.

16. U.S. Senate, Committee on Aeronautical and Space Sciences, *Project Mercury,* 19.

17. Ibid., 51.

18. On connections between *Star Trek* and spaceflight, see Day, "Boldly Going."

19. Rose, *The Politics of Life Itself.*

20. Foucault, *Politics, Philosophy, Culture,* 257.

21. Canguilhem, "The Living and Its Milieu," 7.

22. Ibid.

23. On the first of these two projects, see Van de Vijver, Van Speybroeck, and De Waele, "Epigenetics"; Frank et al., "Interactive Role of Genes and the Environment"; Lock, "The Epigenome and Nature/Nurture Reunification";

Landecker, *Culturing Life.* On the second, see Rose, *The Politics of Life Itself*; Paxson, "Post-Pasteurian Cultures"; Helmreich, *Alien Ocean.*

24. Wynter and McKittrick, "Unparalleled Catastrophe for Our Species?," 37.

25. See Clarke et al., "Biomedicalization."

26. Kotarba, "Social Control Function of Holistic Health Care," 279.

27. Ibid., 285.

28. Institute of Medicine, *Health Standards for Long Duration and Exploration Spaceflight.*

29. Institute of Medicine, *Safe Passage.*

30. Pitts, *The Human Factor.*

31. See Clynes and Kline, "Cyborgs and Space."

32. See, for example, Institute of Medicine, *Safe Passage.*

33. NASA, "Man-Systems Integration Standards," sec. 5.9.2.

34. Ibid., sec. 5.1.2.3.2.

35. Foucault, *The Birth of the Clinic*; Redfield, *Space in the Tropics*; Stepan, *"The Hour of Eugenics"*; Nash, *Inescapable Ecologies.*

36. Penley, *NASA/Trek.*

37. Casper and Moore, "Inscribing Bodies." NASA institutional mission management culture is becoming less hegemonically definitive of space exploration gender and sexuality; NASA people actively participate in commercial space activities and virtual outer space world making in which alternative discourses shape other kinds of spacefaring bodies and social practices.

38. Ackmann, *The Mercury 13*; Weitekamp, *Right Stuff, Wrong Sex*; Tanya Lee Stone, *Almost Astronauts.*

39. Martin, *Flexible Bodies*; Hogle, "Enhancement Technologies and the Body," 695. See also Clarke et al., "Biomedicalization"; Elliott, *Better Than Well.*

40. Mission statement of the Space Life Sciences Directorate, 2008, https://sservi.nasa.gov/articles/space-life-sciences-directorate.

41. Keating and Cambrosio, "Biomedical Platforms," 354.

42. Ibid., 353, quoting *Stedman's Medical Dictionary.*

43. Foucault, *The Birth of the Clinic*; Jordanova, "Earth Science and Environmental Medicine"; Nash, *Inescapable Ecologies.*

44. Janković, *Confronting the Climate*; Foucault, *Discipline and Punish*; Rabinow, *French Modern*; Stepan, *Picturing Tropical Nature*; Redfield, *Space in the Tropics.*

45. Institute of Medicine, *Environmental Medicine.*

46. Aerospace Medical Association, "About AsMA: Overview: What Is Aerospace Medicine?," accessed November 7, 2017, https://www.asma.org.

47. Lovelock, *Gaia.*

48. An anthropological version of this evolutionary model can be found in the writings of anthropologist Ben Finney and astronomer Eric Jones, two advocates of spacefaring who hold that space colonization will foster human diversity by enabling speciation in a cosmic multiplicity of environments. Finney and Jones, *Interstellar Migration and the Human Experience.*

49. Compare Kyle's observation that "Earth" is a misnomer to a microbial oceanographer's declaration that Earth should be called "Ocean" or "Ocean Life." See Helmreich, *Alien Ocean,* 3. As Helmreich notes, this microbial oceanographer uses space satellite imagery of the whole Earth to represent his work as "reaching across scales of human and planetary embodiment," a goal and perspective that demands, as it does for space life scientists, a deliberate rethinking of how environments and ecologies are named, understood, and represented. Ibid., 4.

50. Hersch, *Inventing the American Astronaut,* 138.

51. Collier and Lakoff, "Vital Systems Security."

52. See Conley, *Ecopolitics*; Strydom, *Risk, Environment, and Society*; Petryna, *Life Exposed*; Shostak, "Environmental Justice and Genomics"; Frickel, *Chemical Consequences.*

53. Foucault, *The Birth of the Clinic,* 199. Some explorations of this topic make use of Canguilhem's historical analysis of how scientific medicine came to define "normality" in terms of statistical values and of Foucault's work on how that scientific project supported ways to effect those values in bodies and populations.

54. NASA, *Bioastronautics Roadmap,* 1.

55. Waligora, Horrigan, and Nocogossian, "The Physiology of Spacecraft," 171.

56. Finkelstein, *Attitudes and Disabled People,* 22.

57. Michael Robinson, a historian of exploration culture, describes ways in which contemporary researchers have attempted to locate "novelty-seeking" and "risk-seeking" behaviors in the heritable or environmental formation of personalities linked to individual genetic proclivities, such as having a "longer sequence in the D4 dopamine receptor gene." Robinson, "The Explorer Gene." None of my interlocutors referenced this research, but they created discursive links between attributing "exploratory behavior" to genetic expression and arguments that maintaining a national environment of support for exploration is critical to the perpetuation of an American national "explorer" character (see chapter 1). Robinson cites this as a genetic dimension of American exceptionalism. This can also be interpreted as a modulation of molecular biopolitics, but one in which national environment is explicitly invoked as a necessary genetic "activator." See Rose, *The Politics of Life Itself.*

58. Canguilhem, *The Normal and the Pathological,* 171, 201.

59. See Hacking, "Making Up People."

60. Mol, *The Body Multiple.*

61. Unavailable at the time of this writing, but integrated into headers on the AIAA website, http://www.aiaa.org.

62. Rhatigan, Charles, and Edwards, "Exploration Health Risks," 1.

63. Stamatelatos, *Probabilistic Risk Assessment,* 1.

64. Hersch, "Checklist."

65. Cutting, "Ashes in Orbit," 355. Contemporary notions within astronautics, metaphysical religious, and transhumanist advocacy networks that portray outer space as a transcendentally vital domain in which it will become possible to avoid extinction and connect with transcendental notions of universal consciousness such as the "noosphere" are genealogically related to the philosophical writings and theoretical astronautical calculations of Konstantin Tsiolkovsky, the originator of the still-used Tsiolkovsky rocket equation and a follower of Russian cosmism, a movement to instantiate the next cosmically directed phases of human evolution.

66. Meadows, *Thinking in Systems,* 12.

67. Foucault's use of the system descriptor has not been thoroughly analyzed, especially in relation to his theorizations of discursive formations and their shifts. See Foucault, *The Birth of Biopolitics.*

68. Agamben, *Homo Sacer.*

69. Lock, *Twice Dead.*

70. Cazdyn, *The Already Dead.*

71. Vaughan, *The* Challenger *Launch Decision.*

72. Stepaniak and Lane, *Loss of Signal,* 97.

73. Ibid., 93.

74. In addition to writing this part of the report, Jonathan Clark has done much of the work to formalize "crew survivability" as a program, including one that has a spaceflight patch logo.

75. NASA, "Twins Study: About," accessed June 1, 2017, https://www.nasa.gov.

76. NASA, "NASA's Twins Study Explores Space through You: Videos Highlight Omics," April 20, 2016, https://www.nasa.gov.

## 3. Separation

1. Strathern, "Reading Relations Backwards." See also this book's Introduction.

2. See Haraway, "Situated Knowledges"; Strathern, *Partial Connections*; Merlan, "Male–Female Separation and Forms of Society."

3. Pedersen, "The Fetish of Connectivity."

4. Fortun, "From Latour to Late Industrialism."

5. Ibid., 309.

6. See Auyero and Swistun, *Flammable*; Little, *Toxic Town*; Checker, *Polluted Promises*; Mitman, *Breathing Space*; Masco, *The Nuclear Borderlands.*

7. Strathern, *Partial Connections,* 111.

8. For discussion of such separation and dividing tools and objects, see Kockelman, "The Anthropology of an Equation"; Foucault, *The Birth of the Clinic*; Pritchard, *Confluence*; Sherif, *The American Wall.*

9. Griffin, "Systems Engineering and the 'Two Cultures' of Engineering."

10. NASA, *Suited for Spacewalking.*

11. On the iconic status of the Apollo-era astronaut, see Launius, "Heroes in a Vacuum."

12. Hersch, "High Fashion."

13. Shaw, "Bodies Out of This World."

14. De Monchaux, *Spacesuit,* 3.

15. Young, *Spacesuits.*

16. Shaw, "Bodies Out of This World," 142.

17. De Monchaux, *Spacesuit,* 77.

18. See Hogle, "Enhancement Technologies and the Body"; Elliott, *Better Than Well.*

19. Boyer, "The Corporeality of Expertise."

20. Foster, *Integrating Women into the Astronaut Corps,* 116–18.

21. Launius, "Heroes in a Vacuum," 8.

22. Ibid., 9.

23. Pitts et al., *Astronaut Bio-Suit for Exploration Class Missions.*

24. Ibid., abstract, 5.

25. Ibid., sec. 5.2.1.2, emphasis added.

26. Newman et al., *Astronaut Bio-Suit System for Exploration Class Missions,* 2.

27. Harding, *Survival in Space,* 113.

28. Pitts et al., *Astronaut Bio-Suit for Exploration Class Missions,* sec. 1.2.4.1.

29. Jablonski, *Skin*; see especially the chapter "Future Skin," in which Jablonski discusses the reconfiguration of sensory communication.

30. Newman et al., *Astronaut Bio-Suit System for Exploration Class Missions,* 2.

31. Pitts et al., *Astronaut Bio-Suit for Exploration Class Missions,* sec. 1.3.1.

32. Ibid., sec. 3.2.

33. Deleuze and Guattari, *Anti-Oedipus,* 36, 284.

34. Rebecca Herzig's argument that Arctic polar exploration thrived on an ethos of willing sacrifice and voluntary privation is compelling within the historical framework she puts on it (late nineteenth and early twentieth centuries), but it does not explain the visionary trajectories of extreme environment exploration as both an act of endurance and a challenge to maximize the pleasures of risk taking and the sensual and affective possibilities of *bio-* and *eco-*prefixed design modalities. Herzig, *Suffering for Science,* esp. chap. 3.

35. Latour, "Spheres and Networks," 6.

36. Howes, "Hyperesthesia," 286.

37. Saxon, "Strawberry Fields as Extreme Environments."

38. Mendell, "A Gateway for Human Exploration of Space?," 14.

39. SSP 2006 Team, *Luna Gaia.*

40. See Bass, Wales, and Shalin, "Choosing Mars-Time"; Shirley, *Managing Martians.*

41. Meltzer, *When Biospheres Collide.*

42. Helmreich expands on his arguments in *Alien Ocean* in Helmreich, "The Signature of Life"; and Helmreich, "Extraterrestrial Relativism."

43. An example of these considerations during Constellation can be seen in the minutes of the meeting of the NASA Advisory Committee's Planetary Protection Subcommittee, November 6–7, 2008.

44. Daly and Frodeman, "Separated at Birth, Signs of Rapprochement."

45. Ibid., 147.

46. Golley, "Environmental Ethics and Extraterrestrial Ecosystems."

47. The foundational collection on extraterrestrial ethics remains Hargrove's 1986 volume *Beyond Spaceship Earth.* A conference at NASA Ames in 2007 yielded papers slated for an updated edition of that work, but it has not yet been published.

48. Churchman, *The Systems Approach,* 183–84.

49. The testing at White Sands is described on the facility's website, accessed April 29, 2017, https://www.nasa.gov/centers/wstf/index_new.html.

50. For histories of JPL at different social registers, see Pendle, *Strange Angel*; Conway, *Exploration and Engineering*; Steltzner and Patrick, *The Right Kind of Crazy.*

51. On the Voyagers, see Helmreich, "Remixing the Voyager Interstellar Record."

52. Peter Smith, quoted in David, "Toxic Mars."

53. McKay and Marinova, "The Physics, Biology, and Environmental Ethics."

54. McKay, "Planetary Ecosynthesis on Mars."

55. *Space Odyssey.*

## 4. Transhabitation

1. On prototypes as artifacts that perform ideas in their formative stages, see Suchman et al., "Reconstructing Technologies as Social Practice"; Suchman, Blomberg, and Trigg, "Working Artefacts." On how scenario building in human–computer interactions makes "concretizable" stories about the nature of activities, uses, and functions, see Carroll, *Making Use.* On the imbrication of ideas of healthiness into health care identification technologies, see Godin, "The Rhetoric of a Health Technology." On how filmic scenario-based conceptualizations of prototechnologies and normalized uses (which Kirby calls "diegetic prototypes") become catalysts for the making of actual prototype objects, see Kirby, "The Future Is Now."

2. Adams, "Interview with Constance Adams," 221.

3. Kennedy, "Vernacular of Space Architecture."

4. See Kennedy, "Lessons from TransHab," 2.

5. Escobar, "Sustainability."

6. Ibid., 138.

7. Ibid., 139.

8. Murphy, "Design and Anthropology," 439, 444.

9. See Valentine, "Atmosphere."

10. Anker, "The Ecological Colonization of Space." See also Redfield, "The Half-Life of Empire in Outer Space."

11. Dethloff, *Suddenly, Tomorrow Came . . . .*

12. Adams, "Interview with Constance Adams," 213.

13. Simondon, *On the Mode of Existence of Technical Objects.*

14. For a historical analysis of Faget's thought and processes, see Andrew Chaiken, "How the Spaceship Got Its Shape."

15. Adams, "Interview with Constance Adams," 225.

16. Hitt, Garriott, and Kerwin, *Homesteading Space,* 27.

17. The description of Wernher von Braun as the architect of the U.S. space program reflects Cold War references to the head of the Soviet space agency, whose name was unknown, as "the Architect."

18. Reider, *Dreaming the Biosphere,* 88.

19. McCray, *The Visioneers.*

20. Fuller, *Earth, Inc.*

21. See Gitelson et al., *Manmade Closed Ecological Systems,* 368.

22. Paine, "Biospheres and Solar System Exploration."

23. Murphy, *Swedish Design,* 33.

24. Kennedy, "Lessons from TransHab," 10.

25. Ibid.; Kennedy, "TransHab Project," 81.

26. Editors' introduction to Kennedy, "TransHab Project," 81.

27. Deleuze and Guattari, *A Thousand Plateaus,* 476.

28. See NASA, "Space Settlements," accessed November 8, 2017, https://settlement.arc.nasa.gov.

29. Penley, *NASA/Trek.*

30. A detailed description of this schema, which I heard described in a conference venue, can be found in Grierson, "Beyond NASA."

31. See Wilson, "Problems Common to the Fields of Astronomy and Biology." Notorious NASA and ex-Nazi aerospace "father of space medicine" Hubertus Strughold wrote the keynote review of these shared problems for the symposium in which Wilson's introductory piece appeared. See Strughold, "General Review of Problems Common to the Fields of Astronomy and Biology."

32. See Rabinow, *French Modern*; Fennell, *Last Project Standing.*

33. Kennedy, "Lessons from TransHab," 10.

34. Von Bengtson and Twyford, "Human Subsystem."

35. Stewart, *A Space on the Side of the Road,* 179–81.

36. Haraway, *Simians, Cyborgs, and Women,* 211.

37. Elden, "Secure the Volume," 49.

38. NASA, "Man-Systems Integration Standards," sec. 8.6.3.2, emphasis added.

39. Ibid., sec. 8.6.2.3.

40. Ibid., secs. 8.6.2.3, 8.6.2.4.

41. Ibid., sec. 8.6.2.4.

42. Jones, "System Symptoms."

43. Constance Adams, quoted at Design Feast, "Series: Quotes," accessed November 8, 2017, http://www.designfeast.com.

44. Vogler, "The Universal House," 77.

45. Ibid.

46. As I indicate in the Introduction to this book, there is a whole critical subgenre of the space age and its failures, which also includes critical design history literature such as Sean Topham's *Where's My Space Age?* Nevertheless, contemporary architecture and design literature and projects are reappropriating "space-age" terms like *spacecraft* and *escape vehicle* for critical engagements and experiments with spatial and environmental constraints, modularity, isolation, and "hideouts" or "escape." See Klanten and Feireiss, *Spacecraft*; Morsiani and Smith, *Andrea Zittel.*

47. Armstrong, "Space Is an Ecology for Living In," 128.

48. Sadler, *Archigram,* 100. Archigram's and Deleuze and Guattari's "nomads" are not completely compatible, nor is it clear how they were related historically.

49. Cook and Archigram, *Archigram,* 74–85.
50. Ibid., 80.
51. Simondon, *On the Mode of Existence of Technical Objects,* 59.
52. Sauvagnargues, "Crystals and Membranes."
53. Shapiro, "Attuning to the Chemosphere."
54. Ibid., 369.
55. For a discussion of how Simondon's theorization of individuation and Deleuze's membrane and fold concepts share a view of interiors as always selectively defined by exterior-engaged edges, see Sauvagnargues, "Crystals and Membranes," 69–70.
56. Kallipoliti, "Closed Worlds."
57. Cohen, Flynn, and Matossian, "Water Walls Architecture."
58. Vogler and Vittori, "Space Architecture for the Mother Ship," 393.
59. Larson and Pranke, *Human Spaceflight,* 539–40.
60. Solan, "'Built for Health.'"
61. Adams, "Sociokinetic Analysis as a Tool," 174.
62. Whitehead, "Garrett Finney Brings Space Habitability Down to Earth."
63. Chris, "Faro Cricket."
64. Vogler, "The Universal House," 78.
65. Garrett Finney, quoted in Chris, "Faro Cricket."
66. Garrett Finney, quoted in Rus and Frieze, "Garrett Finney," 138.
67. Gray, "Making Some Connections," 1.
68. Garrett Finney, quoted in Whitehead, "Garrett Finney Brings Space Habitability Down to Earth," 208.

## 5. The Solar Ecosystem

1. Another frequently used label is near-Earth asteroid, or NEA, but the NEO categorization includes cometary and other objects, including technologies. NEO is the term chosen by my JSC interlocutors during their activist work, so I retain it here. Material in this chapter first appeared in Olson, "Political Ecology in the Extreme" and Olson, "NEOspace."
2. It is also known as the K–T boundary.
3. Star and Griesemer, "Institutional Ecology, 'Translations,' and Boundary Objects." See also the introduction to a special issue of *Revue d'Anthropologie des Connaissances* that revisits the "boundary object" concept as a way to understand shared objects within environmental management issues, in which Pascale Trompette calls attention to the concept's genesis as a response to the delineation of "ecologies" more broadly thought of as "worlds" than as institutions. Trompette, "Revisiting the Notion of Boundary Object." However, Trompette

and the other contributors to this special issue are still focused on the ecological properties of objects at work, and not on the constitution of ecologies qua ecologies. Nevertheless, Trompette usefully points out that Star was a student of Anselm Strauss and as such is interested in the relational creation and negotiation of social environments. This legacy reflects Strauss's importation of ecological metaphors to create "social worlds/arenas theory" as a collectivist and interactionist response to positivist organizational theory, which also has the effect of operationalizing "environment" in a way that does not encourage describing it "on its own terms." See Clarke, "Social Worlds/Arenas Theory as Organizational Theory," 124. This is the unproblematized notion of environment and ecology at work in Star and Griesemer's analysis, but their work anticipates contemporary social scientific references to "ecologies" as spatialized sets of interactive and organizationally focused processes.

4. Nussbaum, "Objectification," 388.

5. Barad, *Meeting the Universe Halfway.*

6. McLean, "Black Goo."

7. Ibid., 592.

8. Popular histories of the solar system tend to focus on it as planetary space, discussing the role of the asteroid belt in new calculations of solar systemic relations. See Chambers and Mitton, *From Dust to Life*; Sobel *The Planets*; Corfield, *Lives of the Planets.*

9. Povinelli, *Geontologies.*

10. Stewart and Harding, "Bad Endings."

11. I take this self-designation from e-mail salutations; some members also refer to themselves as "NEOphytes." "NEOphiles" is an informal designation that not all members of the network use, nor is it a designation that the group uses for any official purpose. However, it provides a convenient way to refer to the group/network and its members, and it expresses their commonly held passion for NEOs.

12. On the making of modern American "fear ecologies," cocreated by environmental crises, development, and media of millennial imaginaries, see Davis, *Ecology of Fear.*

13. See Douglas, *Purity and Danger.*

14. Barnes-Svarney, *Asteroid.*

15. I calculated the total number of small bodies by selecting all asteroid and comet classes in the options offered by the JPL Small-Body Database Search Engine, accessed December 21, 2017, https://ssd.jpl.nasa.gov.

16. Galison, *Image and Logic.*

17. Alvarez et al., "Extraterrestrial Cause for the Cretaceous-Tertiary Extinction."

18. Ibid., 1095.

19. For a discussion of nuclear bombs as objects that remade modern cosmology and rescaled imaginaries of environmental security regimes, see Masco, *The Nuclear Borderlands.*

20. Dave Morrison, developer of the Ames Research Center's projects concerning asteroid and comet impact hazards, reports on this meeting in a news article archived online, "Edward Teller (1908–2003) and Defense against Asteroids," October 9, 2003, http://archive.li/f29Is. Morrison indicates that these notes on "Teller at Los Alamos, January 1992" are "adapted from an unpublished book manuscript." When Gene Shoemaker started work for the U.S. Geological Survey in the late 1950s, he compared the structure and mechanics of craters caused by meteorites with those caused by nuclear explosions. Dedicated early on to the idea that impact was a fundamental force in solar system evolution, Shoemaker founded the 1973 Planet-Crossing Asteroid Survey at Palomar Observatory, which would become the model for tracking NEOs.

21. See Mellor, "Colliding Worlds." Mellor discusses the ways that military cooperation in asteroid research legitimates the idea of war in space and the development of space-based weapons and platforms for testing systems such as the Strategic Defense Initiative, which are otherwise opposed politically. Certainly, these proposals bring space scientist and military groups together and perpetuate "the politics of fear" at the heart of U.S. defense policy. However, as I note below, there is also scientific and political pushback against the utility of such weapons. Proposals to develop nonweaponized NEO mitigation strategies, and to secure international collaboration to do so, provide counternarratives that also call for greater understanding of the impacts of technology on the environment.

22. Garretson and Kaupa, "Planetary Defense," 1.

23. Morrison, *The Spaceguard Survey.*

24. This phrase is used in the 2010 U.S. space policy to include both natural and "man-made" threats. See "National Space Policy of the United States of America," White House, June 28, 2010, https://obamawhitehouse.archives.gov.

25. Stephen Ostro, interview in *Planetary Defense.* Clarke's imagined asteroid-watch program appears in his novel *Rendezvous with Rama.*

26. Morrison, "The Impact Hazard," 163.

27. Ibid.

28. Bobrowsky and Rickman, *Comet/Asteroid Impacts and Human Society,* vi.

29. Beck, *Risk Society.*

30. See Stern, "Myriad Planets in Our Solar System."

31. Ingold, "Globes and Spheres."

32. See Sloterdijk, *In the World Interior of Capital.* In his treatments of the relationship between authoritative constructions of spherical interiority and exteriority and modern worlds, Sloterdijk relates the spherical construction of world systems to early modern astronomical cosmic spheres and the development of space stations.

33. Masco, *The Nuclear Borderlands,* 43–98.

34. Yeomans, "The Torino Scale Shows We're Safe."

35. Jasanoff, "Ordering Knowledge, Ordering Society."

36. Association of Space Explorers, "About ASE," accessed November 27, 2017, http://www.space-explorers.org.

37. Collier and Lakoff, "The Vulnerability of Vital Systems," 30. On the ASE petition, see Schweickart et al., *Asteroid Threats.*

38. B612 Foundation website, accessed December 12, 2007, https://b612foundation.org.

39. Ibid.

40. Russell Schweickart, testimony before the Space and Aeronautics Subcommittee of the House Committee on Science and Technology, Washington, D.C., October 11, 2007.

41. Boyle, "Flying up to Meet Asteroids."

42. Schweickart et al., *Asteroid Threats.*

43. Ibid., 46.

44. Ibid., 14.

45. Ibid., 49.

46. Asteroid Day, "Our Story" and "Video," accessed December 24, 2017, https://asteroidday.org.

47. Daston, "Objectivity and the Escape from Perspective." NEOs in this way close the gap on the "overview effect" and its technical/romantic "objectivity" presented in Chapter 1. As with that example, Daston's essay on the process by which "objectivity" is purged of moral and situated perspectival knowledge provides a guidepost for understanding what technical objectivity has become in official and formal discourses and practices. However, astronautical objectivity, as I have attempted to show, is a combination of idealized objectivity and a "from space" perspective considered to be universally and transcendentally "human" in combined epistemological, ontological, emotional, and moral terms.

48. Korsmeyer, Landis, and Abell, "Into the Beyond." This is the title of the study as presented by David Korsmeyer at the International Astronautical Congress in 2007.

49. Blumenberg, *The Genesis of the Copernican World.*

50. Star and Griesemer, "Institutional Ecology, 'Translations,' and Boundary Objects."

51. Lo, "The Interplanetary Superhighway and the Origins Program," 18.

52. Easterbrook, "The Sky Is Falling"; Klesius, "The Million Mile Mission"; Bormanis, "Worlds Beyond."

53. Latour, *Science in Action*; Epstein, *Impure Science,* 3.

54. Billings, "Overview," 483. See also Siddiqi, "Making Spaceflight Modern"; Bainbridge, *The Spaceflight Revolution.* Bainbridge pioneered the view of spaceflight advocacy as a social movement, but more recently NASA collected perspectives on spaceflight and exploration advocacy in the conference collection that includes the works by Billings and Siddiqi cited here; see Dick and Launius, *Societal Impact of Spaceflight.*

55. Review of U.S. Human Spaceflight Plans Committee, *Seeking a Human Spaceflight Program,* 9, 15.

56. Ibid., 15.

57. Martin, *Flexible Bodies.*

58. Cole, "$50,000,000,000,000 from the Asteroids."

59. Lewis, *Mining the Sky.*

60. On bioprospecting, see Hayden, *When Nature Goes Public.*

61. Mitchell, *Carbon Democracy.*

62. Worster, *Nature's Economy,* 37.

## Conclusion

1. Launius and McCurdy, "Introduction."

2. Betmann, "A Counterspace Awakening?"

3. National Defense Authorization Act for Fiscal Year 2014, December 2013, 113th Congress, 1st Session, HR 3304, Public Law 113-66, printed for the use of the Committee on Armed Services of the House of Representatives, sec. 914, p. 158.

4. Betmann, "A Counterspace Awakening?"

5. "Announcing an Upcoming Conference: 55th Robert H. Goddard Memorial Symposium," posted on Roger Launius's Blog, March 3, 2017, https://launiusr.wordpress.com.

6. See Kaplan, "Trump Signs NASA Bill."

7. See Milman, "Trump to Scrap NASA Climate Research"; Milman, "Paris Climate Deal."

8. Waldman, "NASA Nominee Wants to Study Climate Change."

# BIBLIOGRAPHY

Ackmann, Martha. *The Mercury 13: The Untold Story of Thirteen American Women and the Dream of Space Flight.* New York: Random House, 2003.

Ackoff, Russell Lincoln. *Redesigning the Future: A Systems Approach to Societal Problems.* New York: John Wiley, 1974.

Adams, Constance. "Interview with Constance Adams: We Are in This Thing Together." In *After Taste: Expanded Practice in Interior Design,* edited by Kent Kleinman, Joanna Merwood-Salisbury, and Lois Weinthal, 212–25. New York: Princeton Architectural Press, 2012.

———. "Sociokinetic Analysis as a Tool for Optimization of Environmental Design." In *Isolation: NASA Experiments in Closed-Environment Living,* edited by Helen W. Lane, Richard L. Sauer, and Daniel L. Feeback, 165–75. San Diego, Calif.: Univelt, for American Astronautical Society, 2002.

Addelson, Kathryn Pyne. "The Man of Professional Wisdom." In *Beyond Methodology: Feminist Scholarship as Lived Research,* edited by Mary Margaret Fonow and Judith A. Cook, 16–34. Bloomington: Indiana University Press, 1991.

Agamben, Giorgio. *Homo Sacer: Sovereign Power and Bare Life.* Stanford, Calif.: Stanford University Press, 1998.

Agrawal, Arun. *Environmentality: Technologies of Government and the Making of Subjects.* Durham, N.C.: Duke University Press, 2005.

Alvarez, Luis W., Walter Alvarez, Frank Asaro, and Helen V. Michel. "Extraterrestrial Cause for the Cretaceous-Tertiary Extinction." *Science* 208, no. 4448 (1980): 1095–108.

Anker, Peder. "The Ecological Colonization of Space." *Environmental History* 10, no. 2 (2005): 239–68.

Appel, Hannah. "Offshore Work: Oil, Modularity, and the How of Capitalism in Equatorial Guinea." *American Ethnologist* 39, no. 4 (2012): 692–709.

Appiah, Kwame Anthony. "Is the Post- in Postmodernism the Post- in Postcolonial?" *Critical Inquiry* 17, no. 2 (1991): 336–57.

Arendt, Hannah. *Between Past and Future: Eight Exercises in Political Thought.* New York: Penguin Books, 2006.

Armstrong, Rachel. "Space Is an Ecology for Living In." In *Space Architecture: The New Frontier for Design Research,* edited by Neil Leach, 128–33. London: John Wiley, 2014.

Arnold, Darrell P. "Systems Theory: A Secret History of the Twentieth Century." In *Traditions of Systems Theory: Major Figures and Contemporary Developments,* ed. Darrell P. Arnold, 10–20. New York: Routledge, 2014.

Atwill, William D. *Fire and Power: The American Space Program as Postmodern Narrative.* Athens: University of Georgia Press, 1994.

Auyero, Javier, and Débora Alejandra Swistun. *Flammable: Environmental Suffering in an Argentine Shantytown.* Oxford: Oxford University Press, 2009.

Bainbridge, William Sims. *The Spaceflight Revolution: A Sociological Study.* 1976. Reprint, Malabar, Fla.: Krieger, 1983.

Barad, Karen Michelle. *Meeting the Universe Halfway: Quantum Physics and the Entanglement of Matter and Meaning.* Durham, N.C.: Duke University Press, 2007.

Barnes-Svarney, Patricia L. *Asteroid: Earth Destroyer or New Frontier?* New York: Plenum, 1996.

Bass, Deborah S., Roxana C. Wales, and Valerie L. Shalin. "Choosing Mars-Time: Analysis of the Mars Exploration Rover Experience." Paper presented at the IEEE Aerospace Conference, March 5–12, 2005. http://ntrs.nasa.gov.

Bateson, Gregory. *Steps to an Ecology of Mind.* 1972. Reprint, Chicago: University of Chicago Press, 2000.

Battaglia, Debbora. *E.T. Culture: Anthropology in Outerspaces.* Durham, N.C.: Duke University Press, 2005.

Battaglia, Debbora, David Valentine, and Valerie Olson. "Relational Space: An Earthly Installation." *Cultural Anthropology* 30, no. 2 (2015): 245–56.

Baudrillard, Jean. *Simulacra and Simulation.* Translated by Sheila Faria Glaser. Ann Arbor: University of Michigan Press, 1994.

Bebbington, Anthony, and Jeffrey Bury. *Subterranean Struggles: New Dynamics of Mining, Oil, and Gas in Latin America.* Austin: University of Texas Press, 2013.

Beck, Ulrich. *Risk Society: Towards a New Modernity.* London: Sage, 1992.

Beer, Gillian. *Darwin's Plots: Evolutionary Narrative in Darwin, George Eliot, and Nineteenth-Century Fiction.* London: Routledge & Kegan Paul, 1983.

Benjamin, Marina. *Rocket Dreams: How the Space Age Shaped Our Vision of a World Beyond.* New York: Free Press, 2003.

Betmann, Maximilian. "A Counterspace Awakening? (Part 2): Assessing the Recent Shift in US National Security Space Strategy." The Space Review, May 30, 2017. http://www.thespacereview.

Biagioli, Mario. *Galileo's Instruments of Credit: Telescopes, Images, Secrecy.* Chicago: University of Chicago Press, 2006.

Biersack, Aletta, and James B. Greenberg, eds. *Reimagining Political Ecology.* Durham, N.C.: Duke University Press, 2006.

Billings, Linda. "Overview: Ideology, Advocacy, and Spaceflight—Evolution of a Cultural Narrative." In *Societal Impact of Spaceflight,* edited by Steven J. Dick and Roger D. Launius, 483–99. Washington, D.C.: NASA, 2007.

Blumenberg, Hans. *The Genesis of the Copernican World.* Cambridge: MIT Press, 1987.

Bobrowsky, Peter T., and Hans Rickman, eds. *Comet/Asteroid Impacts and Human Society: An Interdisciplinary Approach.* Berlin: Springer, 2007.

Boellstorff, Tom. *Coming of Age in Second Life: An Anthropologist Explores the Virtually Human.* Princeton, N.J.: Princeton University Press, 2008.

Bormanis, André. "Worlds Beyond." *Planetary Report,* November/December 2007.

Boyer, Dominic. "The Corporeality of Expertise." *Ethnos* 70, no. 2 (1995): 243–66.

Boyer, Dominic, and George E. Marcus. "Introduction: New Methodologies for a Transformed Discipline." In *Theory Can Be More Than It Used to Be: Learning Anthropology's Method in a Time of Transition,* edited by Dominic Boyer, James D. Faubion, and George E. Marcus, 1–12. Ithaca, N.Y.: Cornell University Press, 2015.

Boyle, Rebecca. "Flying up to Meet Asteroids: A Proposed NASA Mission to Intercept an Ill-Omened Rock in the Sky." PopSci: The Future Now (blog). *Popular Science,* March 25, 2009. http://www.popsci.com.

Braun, Bruce. "Environmental Issues: Global Natures in the Space of Assemblage." *Progress in Human Geography* 30, no. 5 (2006): 644–54.

Brill, James H. "Systems Engineering: A Retrospective View." *Systems Engineering* 1, no. 4 (1998): 258–66.

Burrough, Bryan. *Dragonfly: NASA and the Crisis aboard Mir.* New York: HarperCollins, 1998.

Burrows, William E. *This New Ocean: The Story of the First Space Age.* New York: Random House, 1998.

Canguilhem, Georges. "The Living and Its Milieu." Translated by John Savage. *Grey Room,* no. 3 (Spring 2001): 7–31.

———. *The Normal and the Pathological.* New York: Zone Books, 1989.

Carrier, James G., and Paige West, eds. *Virtualism, Governance and Practice: Vision and Execution in Environmental Conservation.* New York: Berghahn Books, 2009.

Carrington, Damian. "The Anthropocene Epoch: Scientists Declare Dawn of Human-Influenced Age." *Guardian,* August 29, 2016. https://www.theguardian.com.

Carroll, John M. *Making Use: Scenario-Based Design of Human–Computer Interactions.* Cambridge: MIT Press, 2000.

Carson, Rachel. *Silent Spring.* 40th anniversary ed. Boston: Houghton Mifflin, 2002.

Casper, Monica J., and Lisa J. Moore. "Inscribing Bodies, Inscribing the Future: Gender, Sex, and Reproduction in Outer Space." *Sociological Perspectives* 38, no. 2 (1995): 311–33.

Cazdyn, Eric. *The Already Dead: The New Time of Politics, Culture, and Illness.* Durham, N.C.: Duke University Press, 2012.

Chaiken, Andrew. "How the Spaceship Got Its Shape." *Air & Space,* November 2009. http://www.airspacemag.com.

Chambers, John, and Jacqueline Mitton. *From Dust to Life: The Origin and Evolution of Our Solar System.* Princeton, N.J.: Princeton University Press, 2017.

Chandler, Nahum Dimitri. *X—The Problem of the Negro as a Problem for Thought.* New York: Fordham University Press, 2014.

Chapman, Richard L. *Project Management in NASA: The System and the Men.* Washington, D.C.: NASA, 1973.

Checker, Melissa. *Polluted Promises: Environmental Racism and the Search for Justice in a Southern Town.* New York: New York University Press, 2005.

Checkland, Peter. *Systems Thinking, Systems Practice.* Chichester: John Wiley, 1981.

Choy, Timothy K. *Ecologies of Comparison: An Ethnography of Endangerment in Hong Kong.* Durham, N.C.: Duke University Press, 2011.

Chris. "Faro Cricket: Inner Space." Squob, September 10, 2008. http://squob.com.

Churchman, C. West. *The Systems Approach.* Rev. ed. New York: Dell, 1979.

———. *The Systems Approach and Its Enemies.* New York: Basic Books, 1979.

Clark, Nigel. *Inhuman Nature: Sociable Life on a Dynamic Planet.* London: Sage, 2011.

Clark, Timothy. "Derangements of Scale." In *Telemorphosis: Theory in the Era of Climate Change,* vol. 1, edited by Tom Cohen, 148–66. Ann Arbor, Mich.: Open Humanities Press, 2012.

Clarke, Adele E. "Social Worlds/Arenas Theory as Organizational Theory." In *Social Organization and Social Process: Essays in Honor of Anselm Strauss,* edited by David R. Maines, 119–58. New York: Aldine de Gruyter, 1991.

Clarke, Adele E., Janet K. Shim, Laura Mamo, Jennifer Ruth Fosket, and Jennifer R. Fishman. "Biomedicalization: Technoscientific Transformations of Health, Illness, and U.S. Biomedicine." *American Sociological Review* 68, no. 2 (2003): 161–94.

Clarke, Arthur C. *Rendezvous with Rama.* London: Pan Books, 1973.

Clynes, Manfred E., and Nathan S. Kline. "Cyborgs and Space." In *The Cyborg Handbook,* edited by Chris Hables Gray, 29–34. New York: Routledge, 1995.

Cohen, Marc M., Michael T. Flynn, and Renée Matossian. "Water Walls Architecture: Massively Redundant and Highly Reliable Life Support for Long Duration Exploration Missions." Paper presented at the Global Space Exploration Conference, Washington, D.C., 2012. http://www.spacearchitect.org/pubs.

Cole, Dandridge. "$50,000,000,000,000 from the Asteroids." *Space World* 4, no. 2 (1963): 1–8.

Collier, Stephen J., and Andrew Lakoff. "Vital Systems Security: Reflexive Biopolitics and the Government of Emergency." *Theory, Culture & Society* 32, no. 2 (2015): 19–51.

———. "The Vulnerability of Vital Systems: How 'Critical Infrastructure' Became a Security Problem." In *Securing "the Homeland": Critical Infrastructure, Risk, and (In)Security*, edited by Myriam Dunn Cavelty and Kristian Søby Kristensen. London: Routledge, 2008.

Conley, Verena Andermatt. *Ecopolitics: The Environment in Poststructuralist Thought.* London: Routledge, 1997.

Conway, Erik M. *Exploration and Engineering: The Jet Propulsion Laboratory and the Quest for Mars.* Baltimore: Johns Hopkins University Press, 2015.

Cook, Peter, and Archigram. *Archigram.* New York: Praeger, 1973.

Cooper, Henry S. F., Jr. *Before Lift-Off: The Making of a Space Shuttle Crew.* Baltimore: Johns Hopkins University Press, 1987.

Corfield, Richard. *Lives of the Planets: A Natural History of the Solar System.* New York: Basic Books, 2007.

Cosgrove, Denis E. *Apollo's Eye: A Cartographic Genealogy of the Earth in the Western Imagination.* Baltimore: Johns Hopkins University Press, 2001.

Cutting, Andrew. "Ashes in Orbit: Celestis Spaceflights and the Invention of Post-cremationist Afterlives." *Science as Culture* 18, no. 3 (2009): 355–69.

Daly, Erin Moore, and Robert Frodeman. "Separated at Birth, Signs of Rapprochement: Environmental Ethics and Space Exploration." *Ethics and the Environment* 13, no. 1 (2008): 135–51.

Das, Veena. "The Signature of the State: The Paradox of Illegibility." In *Anthropology in the Margins of the State,* edited by Veena Das and Deborah Poole, 225–52. Santa Fe, N.M.: School of American Research Press, 2004.

Daston, Lorraine. "Objectivity and the Escape from Perspective." *Social Studies of Science* 22 (1992): 597–618.

David, Leonard. "Toxic Mars: Astronauts Must Deal with Perchlorate on the Red Planet." Space.com, June 13, 2013. http://www.space.com.

Davis, Mike. *Ecology of Fear: Los Angeles and the Imagination of Disaster.* New York: Metropolitan Books, 1998.

Day, Dwayne A. "Boldly Going: *Star Trek* and Spaceflight." The Space Review, November 28, 2005. http://www.thespacereview.com.

Dean, Margaret Lazarus. *Leaving Orbit: Notes from the Last Days of American Spaceflight.* Minneapolis: Graywolf Press, 2015.

Deleuze, Gilles. *The Fold: Leibniz and the Baroque.* Translated by Tom Conley. Minneapolis: University of Minnesota Press, 1993.

Deleuze, Gilles, and Félix Guattari. *Anti-Oedipus: Capitalism and Schizophrenia.* Translated by Robert Hurley, Mark Seem, and Helen R. Lane. Minneapolis: University of Minnesota Press, 1983.

———. *A Thousand Plateaus: Capitalism and Schizophrenia.* Translated by Brian Massumi. London: Athlone Press, 1988.

DeLoughrey, Elizabeth. "Satellite Planetarity and the Ends of the Earth." *Public Culture* 26, no. 2 (2014): 257–80.

de Monchaux, Nicholas. *Spacesuit: Fashioning Apollo.* Cambridge: MIT Press, 2011.

Dethloff, Henry C. *Suddenly, Tomorrow Came . . . : A History of the Johnson Space Center.* Washington, D.C.: NASA, 1993.

Dick, Steven J., and Roger D. Launius, eds. *Societal Impact of Spaceflight.* Washington, D.C.: NASA, 2007.

Diprose, Rosalyn. "Biopolitical Technologies of Prevention." *Health Sociology Review* 17 (2008): 141–50.

DNews. "Finding NEEMO: NASA's Asteroid Mission in the Sea." Discovery Channel, September 21, 2011. http://news.discovery.com.

Doel, Ronald Edmund. *Solar System Astronomy in America: Communities, Patronage, and Interdisciplinary Science, 1920–1960.* Cambridge: Cambridge University Press, 1996.

Douglas, Mary. *Purity and Danger: An Analysis of Concepts of Pollution and Taboo.* London: Routledge, 1966.

Dyson, Freeman. "Introduction." In *The High Frontier: Human Colonies in Space,* 3rd ed., edited by Gerard K. O'Neill, 5–7. Burlington, Ont.: Apogee Books, 2000.

Easterbrook, Gregg. "The Sky Is Falling." *Atlantic,* June 2008, 74–84. Reprinted in *The Best American Science Writing, 2009,* edited by Natalie Angier, 288–305. New York: Harper Perennial, 2009.

Easterling, Keller. *Extrastatecraft: The Power of Infrastructure Space.* New York: Verso, 2014.

Elden, Stuart. "Secure the Volume: Vertical Geopolitics and the Depth of Power." *Political Geography* 34 (2013): 35–51.

Elliott, Carl. *Better Than Well: American Medicine Meets the American Dream.* New York: W. W. Norton, 2003.

Epstein, Steven. *Impure Science: AIDS, Activism, and the Politics of Knowledge.* Berkeley: University of California Press, 1996.

Escobar, Arturo. "After Nature: Steps to an Anti-essentialist Political Ecology." *Current Anthropology* 40, no. 1 (1999): 1–16.

———. "Sustainability: Design for the Pluriverse." *Development* 54, no. 2 (2011): 137–40.

Faber, Roland. "Whitehead at Infinite Speed: Deconstructing System as Event." In *Schleiermacher and Whitehead: Open Systems in Dialogue,* edited by Christine Helmer, 39–72. Berlin: Walter de Gruyter, 2004.

Fabian, Johannes. *Time and the Other: How Anthropology Makes Its Object.* Reprint ed. New York: Columbia University Press, 2002.

Farman, Abou. "Speculative Matter: Secular Bodies, Minds, and Persons." *Cultural Anthropology* 28, no. 4 (2013): 737–59.

Farr, Rebecca A., David L. Christensen, and Edward L. Keith. "The Business Case for Spiral Development in Heavy-Lift Launch Vehicle Systems." In *41st AIAA/ASME/SAE/ASEE Joint Propulsion Conference & Exhibit, 10–13 July 2005, Tucson, Arizona.* Reston, Va.: American Institute of Aeronautics and Astronautics, 2005. https://ntrs.nasa.gov.

Fennell, Catherine. *Last Project Standing: Civics and Sympathy in Post-welfare Chicago.* Minneapolis: University of Minnesota Press, 2015.

Finkelstein, Victor. *Attitudes and Disabled People: Issues for Discussion.* New York: International Exchange of Information in Rehabilitation, 1980.

Finney, Ben R., and Eric M. Jones. *Interstellar Migration and the Human Experience.* Berkeley: University of California Press, 1985.

Fortun, Kim. "From Latour to Late Industrialism." *HAU: Journal of Ethnographic Theory* 4, no. 1 (2014): 309–29.

Foss, Sonja K., and Cindy L. Griffin. "Beyond Persuasion: A Proposal for an Invitational Rhetoric." *Communication Monographs* 62 (1995): 1–18.

Foster, Amy E. *Integrating Women into the Astronaut Corps: Politics and Logistics at NASA, 1972–2004.* Baltimore: Johns Hopkins University Press, 2011.

Foucault, Michel. *The Birth of Biopolitics: Lectures at the Collège de France, 1978–1979.* Edited by Michel Senellart. Translated by Graham Burchell. New York: Palgrave Macmillan, 2008.

———. *The Birth of the Clinic: An Archaeology of Medical Perception.* New York: Vintage Books, 1994.

———. *Discipline and Punish: The Birth of the Prison.* New York: Pantheon Books, 1977.

———. "Of Other Spaces." *Diacritics* 16 (1986): 22–27.

———. *Politics, Philosophy, Culture: Interviews and Other Writings, 1977–1984.* Edited by Lawrence D. Kritzman. New York: Routledge, 1988.

———. *Technologies of the Self: A Seminar with Michel Foucault.* Edited by Luther H. Martin, Huck Gutman, and Patrick H. Hutton. Amherst: University of Massachusetts Press, 1988.

Frank, John, Geoffrey Lomax, Patricia Baird, and Margaret Lock. "Interactive Role of Genes and the Environment." In *Healthier Societies: From Analysis to Action,* edited by Jody Heymann, Clyde Hertzman, Morris L. Barer, and Robert G. Evans, 11–34. Oxford: Oxford University Press, 2006.

Franklin, Sarah. *Dolly Mixtures: The Remaking of Genealogy.* Durham, N.C.: Duke University Press, 2007.

Franks, Paul W. *All or Nothing: Systematicity, Transcendental Arguments, and Skepticism in German Idealism.* Cambridge, Mass.: Harvard University Press, 2005.

Frickel, Scott. *Chemical Consequences: Environmental Mutagens, Scientist Activism, and the Rise of Genetic Toxicology.* New Brunswick, N.J.: Rutgers University Press, 2004.

Fuller, R. Buckminster. *Earth, Inc.* Garden City, N.Y.: Anchor Press, 1973.

Gadamer, Hans-Georg. *Wahrheit und Methode.* 2nd ed. Tübingen: J. C. B. Mohr, 1965.

Galison, Peter. *Image and Logic: A Material Culture of Microphysics.* Chicago: University of Chicago Press, 1997.

Galloway, Alexander R., and Eugene Thacker. *The Exploit: A Theory of Networks.* Minneapolis: University of Minnesota Press, 2007.

Garb, Yaakov. "The Use and Abuse of the Whole Earth Image." *Whole Earth Review* 45 (1985): 18–25.

Garretson, Peter, and Douglas Kaupa. "Planetary Defense: Potential Department of Defense Mitigation Roles." Paper presented at the IAA Planetary Defense Conference: Protecting Earth from Asteroids, Washington, D.C., 2007.

Geertz, Clifford. “Thick Description: Toward an Interpretive Theory of Culture.” In *The Interpretation of Cultures: Selected Essays,* 3–30. New York: Basic Books, 1973.

Gerhardt, Uta. *The Social Thought of Talcott Parsons: Methodology and American Ethos.* Farnham, England: Ashgate, 2011.

Gertner, Jon. *The Idea Factory: Bell Labs and the Great Age of American Innovation.* New York: Penguin Books, 2012.

Gibson-Graham, J. K. *The End of Capitalism (as We Knew It): A Feminist Critique of Political Economy.* Minneapolis: University of Minnesota Press, 2006.

Gitelson, I. I., G. M. Lisovsky, and R. D. MacElroy. *Manmade Closed Ecological Systems.* London: Taylor & Francis, 2003.

Godin, Benoit. “The Rhetoric of a Health Technology: The Microprocessor Patient Card.” *Social Studies of Science* 27, no. 6 (1997): 865–902.

Golley, Frank B. “Environmental Ethics and Extraterrestrial Ecosystems.” In *Beyond Spaceship Earth: Environmental Ethics and the Solar System,* edited by Eugene C. Hargrove, 211–26. San Francisco: Sierra Club Books, 1986.

———. *A History of the Ecosystem Concept in Ecology: More Than the Sum of the Parts.* New Haven, Conn.: Yale University Press, 1993.

Gray, Chris Hables, ed. *The Cyborg Handbook.* New York: Routledge, 1995.

Gray, Lisa. “Making Some Connections.” *Houston Chronicle,* August 27, 2008, 1–3.

Grierson, Bruce. “Beyond NASA: Dawn of the Next Space Age.” *Popular Science,* April 17, 2004, 69–75.

Griffin, Michael. “The Space Economy.” Lecture delivered at the NASA 50th Anniversary Lecture Series, Washington, D.C., September 17, 2007. https://www.nasa.gov.

———. “Systems Engineering and the ‘Two Cultures’ of Engineering.” Lecture delivered at the Boeing Distinguished Lecture Series, Purdue University, March 28, 2007. https://www.nasa.gov.

Grosz, Elizabeth. *The Nick of Time: Politics, Evolution, and the Untimely.* Durham, N.C.: Duke University Press, 2004.

Gupta, Akhil. *Red Tape: Bureaucracy, Structural Violence, and Poverty in India.* Durham, N.C.: Duke University Press, 2012.

Hacking, Ian. “Making Up People.” In *Reconstructing Individualism: Autonomy, Individuality, and the Self in Western Thought,* edited by Thomas C. Heller, Morton Sosna, and David E. Wellbery, 222–36. Stanford, Calif.: Stanford University Press, 1986.

Hagen, Joel B. *An Entangled Bank: The Origins of Ecosystem Ecology.* New Brunswick, N.J.: Rutgers University Press, 1992.

Hall, Arthur D. *Metasystems Methodology.* New York: Pergamon, 1989.

Hallyn, Fernand, ed. *Metaphor and Analogy in the Sciences.* Dordrecht: Kluwer Academic, 2000.

Hammond, Debora. *The Science of Synthesis: Exploring the Social Implications of General Systems Theory.* Boulder: University Press of Colorado, 2003.

Handberg, Roger. *Reinventing NASA: Human Spaceflight, Bureaucracy, and Politics.* Westport, Conn.: Praeger, 2003.

Haraway, Donna J. "Cyborgs and Symbionts: Living Together in the New World Order." In *The Cyborg Handbook,* edited by Chris Hables Gray, xi–xx. New York: Routledge, 1995.

———. "A Manifesto for Cyborgs: Science, Technology, and Socialist Feminism in the 1980s." *Socialist Review,* no. 80 (1985): 65–108.

———. *Simians, Cyborgs, and Women: The Reinvention of Nature.* New York: Routledge, 1991.

———. "Situated Knowledges: The Science Question in Feminism and the Privilege of Partial Perspective." *Feminist Studies* 14, no. 3 (1988): 575–99.

Haraway, Donna J., and David Harvey. "Nature, Politics, and Possibilities: A Debate and Discussion with David Harvey and Donna Haraway." *Environment and Planning D: Society and Space* 13, no. 5 (1995): 507–27.

Harding, Richard. *Survival in Space: Medical Problems of Manned Spaceflight.* London: Routledge, 1989.

Hargrove, Eugene C., ed. *Beyond Spaceship Earth: Environmental Ethics and the Solar System.* San Francisco: Sierra Club Books, 1986.

Harvey, David. *The Condition of Postmodernity: An Enquiry into the Origins of Cultural Change.* Oxford: Blackwell, 1989.

Hayden, Cori. *When Nature Goes Public: The Making and Unmaking of Bioprospecting in Mexico.* Princeton, N.J.: Princeton University Press, 2000.

Hayles, N. Katherine. *How We Became Posthuman: Virtual Bodies in Cybernetics, Literature, and Informatics.* Chicago: University of Chicago Press, 1999.

Helmer, Christine, ed. *Schleiermacher and Whitehead: Open Systems in Dialogue.* Berlin: Walter de Gruyter, 2004.

Helmreich, Stefan. *Alien Ocean: Anthropological Voyages in Microbial Seas.* Berkeley: University of California Press, 2009.

———. "Extraterrestrial Relativism." *Anthropological Quarterly* 85, no. 4 (2013): 1125–40.

———. "Remixing the Voyager Interstellar Record: Or, As Extraterrestrials Might Listen." *Journal of Sonic Studies* 8 (2014). http://sonicfield.org.

———. "The Signature of Life: Designing the Astrobiological Imagination." *Grey Room,* no. 23 (Spring 2006): 66–95.

Hersch, Matthew H. "Checklist: The Secret Life of Apollo's 'Fourth Crewmember.'" *Sociological Review* 57, no. s1 (2009): 6–24.

———. "High Fashion: The Women's Undergarment Industry and the Foundations of American Spaceflight." *Fashion Theory* 13, no. 3 (2009): 345–70.

———. *Inventing the American Astronaut.* New York: Palgrave Macmillan, 2012.

Herzig, Rebecca M. *Suffering for Science: Reason and Sacrifice in Modern America.* New Brunswick, N.J.: Rutgers University Press, 2005.

Hitt, David, Owen Garriott, and Joe Kerwin. *Homesteading Space: The Skylab Story.* Lincoln: University of Nebraska Press, 2008.

Hogle, Linda F. "Enhancement Technologies and the Body." *Annual Review of Anthropology* 34, no. 1 (2005): 695–716.

Holyoak, Keith J., and Paul Thagard. *Mental Leaps: Analogy in Creative Thought.* Cambridge: MIT Press, 1995.

Hornborg, Alf, and Carole L. Crumley. *The World System and the Earth System: Global Socioenvironmental Change and Sustainability since the Neolithic.* Walnut Creek, Calif.: Left Coast Press, 2007.

Howes, David. "Hyperesthesia, or, The Sensual Logic of Late Capitalism." In *Empire of the Senses: The Sensual Culture Reader,* edited by David Howes, 281–303. Oxford: Berg, 2005.

Howitt, Richard. "Scale as Relation: Musical Metaphors of Geographical Scale." *Area* 30, no. 1 (1998): 49–58.

Hoyningen-Huene, Paul. *Systematicity: The Nature of Science.* New York: Oxford University Press, 2013.

Hughes, Thomas P., and Agatha C. Hughes. "Introduction." In *Systems, Experts, and Computers: The Systems Approach in Management and Engineering, World War II and After,* edited by Agatha C. Hughes and Thomas P. Hughes, 1–26. Cambridge: MIT Press, 2000.

Ingold, Tim. "Globes and Spheres: The Topology of Environmentalism." In *Environmentalism: The View from Anthropology,* edited by Kay Milton, 31–42. London: Routledge, 1993.

Institute of Medicine. *Environmental Medicine: Integrating a Missing Element into Medical Education.* Washington, D.C.: National Academies Press, 1995.

———. *Health Standards for Long Duration and Exploration Spaceflight: Ethics Principles, Responsibilities, and Decision Framework.* Washington, D.C.: National Academies Press, 2014.

Institute of Medicine, Committee on Creating a Vision for Space Medicine during Travel beyond Earth Orbit, Board on Health Sciences Policy. *Safe Passage: Astronaut Care for Exploration Missions.* Edited by John R. Ball and Charles H. Evans Jr. Washington, D.C.: National Academies Press, 2001.

Jablonski, Nina G. *Skin: A Natural History.* Berkeley: University of California Press, 2006.

Janković, Vladimir. *Confronting the Climate: British Airs and the Making of Environmental Medicine.* New York: Palgrave Macmillan, 2010.

Jasanoff, Sheila. "Ordering Knowledge, Ordering Society." In *States of Knowledge: The Co-production of Science and Social Order,* edited by Sheila Jasanoff, 13–45. London: Routledge, 2004.

Jasanoff, Sheila, and Marybeth Long Martello, eds. *Earthly Politics: Local and Global in Environmental Governance.* Cambridge: MIT Press, 2004.

Jenkins, Henry, with Katie Clinton, Ravi Purushotma, Alice J. Robison, and Margaret Weigel. *Confronting the Challenges of Participatory Culture: Media Education for the Twenty-first Century.* Chicago: John D. and Catherine T. MacArthur Foundation, 2006. https://www.macfound.org.

Jensen, Casper Bruun. "Experimenting with Political Materials: Environmental Infrastructures and Ontological Transformations." *Distinktion: Journal of Social Theory* 16, no. 1 (2015): 17–30.

Johnson, Stephen B. *The Secret of Apollo: Systems Management in American and European Space Programs.* Baltimore: Johns Hopkins University Press, 2002.

Jones, Caroline. "System Symptoms: Caroline A. Jones on Jack Burnham's 'Systems Esthetics' (1968)." *International Art Forum* 51, no. 1 (2012): 113–16.

Jordanova, L. J. "Earth Science and Environmental Medicine: The Synthesis of the Late Enlightenment." In *Images of the Earth: Essays in the History of the Environmental Sciences,* edited by L. J. Jordanova and Roy S. Porter, 119–46. Chalfont St. Giles: British Society for the History of Science, 1979.

Kallipoliti, Lydia. "Closed Worlds: The Rise and Fall of Dirty Physiology." *Architectural Theory Review* 20, no. 1 (2015): 67–90.

Kamler, Kenneth. *Surviving the Extremes: A Doctor's Journey to the Limits of Human Endurance.* New York: St. Martin's Press, 2004.

———. "To Practice for Mars, NASA Dives Deep." *Popular Mechanics,* October 30, 2009. http://www.popularmechanics.com.

Kaplan, Sarah. "Trump Signs NASA Bill Aimed at Sending People to Mars." *Washington Post,* March 21, 2017. https://www.washingtonpost.com.

Keating, Peter, and Alberto Cambrosio. "Biomedical Platforms." *Configurations* 8, no. 3 (2000): 337–87.

Kelty, Christopher M. *Two Bits: The Cultural Significance of Free Software.* Durham, N.C.: Duke University Press, 2008.

Kennedy, Kriss. "Lessons from TransHab, an Architect's Experience." Paper presented at the AIAA Space Architecture Symposium, Houston, October 10–11, 2002.

———. "TransHab Project." In *Out of This World: The New Field of Space Architecture,* edited by A. Scott Howe and Brent Sherwood, 81–88. Reston, Va.: American Institute of Aeronautics and Astronautics, 2009.

———. "Vernacular of Space Architecture." In *Out of This World: The New Field of Space Architecture,* edited by A. Scott Howe and Brent Sherwood, 7–21. Reston, Va.: American Institute of Aeronautics and Astronautics, 2009.

Kirby, David. "The Future Is Now: Diegetic Prototypes and the Role of Popular Films in Generating Real-World Technological Development." *Social Studies of Science* 40, no. 1 (2010): 41–70.

Klanten, Robert, and Lukas Feireiss, eds. *Spacecraft: Fleeting Architecture and Hideouts.* Berlin: Die Gestalten Verlag, 2007.

Klerkx, Greg. *Lost in Space: The Fall of NASA and the Dream of a New Space Age.* New York: Pantheon Books, 2004.

Klesius, Michael. "The Million Mile Mission." *Air & Space,* July 2008. http://www.airspacemag.com.

Kockelman, Paul. "The Anthropology of an Equation: Sieves, Spam Filters, Agentive Algorithms, and Ontologies of Transformation." *HAU: Journal of Ethnographic Theory* 3, no. (2013): 33–61.

Korsmeyer, David J., Rob R. Landis, and Paul A. Abell. "Into the Beyond: A Crewed Mission to a Near-Earth Object." Paper presented at the International Astronautical Congress, Hyderabad, India, September 2007.

Kotarba, Joseph. "Social Control Function of Holistic Health Care in Bureaucratic Settings: The Case of Space Medicine." *Journal of Health and Social Behavior* 24, no. 3 (1983): 275–88.

Kroll-Smith, Steve, Stephen R. Couch, and Brent K. Marshall. "Sociology, Extreme Environments and Social Change." *Current Sociology* 45, no. 3 (1997): 1–18.

Lahsen, Myanna. "Transnational Locals: Brazilian Experiences of the Climate Regime." In *Earthly Politics: Local and Global in Environmental Governance,* edited by Sheila Jasanoff and Marybeth Long Martello, 151–72. Cambridge: MIT Press, 2004.

Lambright, W. Henry. "NASA and the Environment: Science in a Political Context." In *Societal Impact of Spaceflight,* edited by Steven J. Dick and Roger D. Launius, 313–30. Washington, D.C.: NASA, 2007.

———. *NASA and the Environment: The Case of Ozone Depletion.* Washington, D.C.: NASA, 2005.

———. "The Political Construction of Space Satellite Technology." *Science, Technology, and Human Values* 19, no. 1 (1994): 47–69.

———. *Why Mars: NASA and the Politics of Space Exploration.* Baltimore: Johns Hopkins University Press, 2014.

Lambright, W. Henry, and Agnes Gereben Schaefer. "The Political Context of Technology Transfer: NASA and the International Space Station." *Comparative Technology Transfer and Society* 2, no. 1 (2004): 1–30.

Landecker, Hannah. *Culturing Life: How Cells Became Technologies.* Cambridge, Mass.: Harvard University Press, 2007.

Lane, K. Maria D. "Mapping the Mars Canal Mania: Cartographic Projection and the Creation of a Popular Icon." *Imago Mundi* 58, no. 2 (2006): 198–211.

Larkin, Brian. "The Politics and Poetics of Infrastructure." *Annual Review of Anthropology* 42, no. 1 (2013): 327–43.

Larson, Wiley J., and Linda K. Pranke, eds. *Human Spaceflight: Mission Analysis and Design.* New York: McGraw-Hill, 2000.

Latour, Bruno. *Aramis, or, The Love of Technology.* Translated by Catherine Porter. Cambridge, Mass.: Harvard University Press, 1996.

———. "Give Me a Laboratory and I Will Raise the World." In *Science Observed: Perspectives on the Social Study of Science,* edited by Karin Knorr-Cetina and Michael Mulkay, 141–70. London: Sage, 1983.

———. *Science in Action: How to Follow Scientists and Engineers through Society.* Cambridge, Mass.: Harvard University Press, 1987.

———. "Spheres and Networks: Two Ways to Reinterpret Globalization." Lecture delivered at the Harvard University Graduate School of Design, February 17, 2009.

Launius, Roger D. "Heroes in a Vacuum: The Apollo Astronaut as Cultural Icon." Paper presented at the 43rd AIAA Aerospace Sciences Meeting and Exhibit, Reno, Nevada, 2005.

———. "Writing the History of Space's Extreme Environment." *Environmental History* 15, no. 3 (2010): 526–32.

Launius, Roger D., and Howard E. McCurdy. "Introduction: The Imperial Presidency in the History of Space Exploration." In *Spaceflight and the Myth of Presidential Leadership,* edited by Roger D. Launius and Howard E. McCurdy, 1–14. Urbana: University of Illinois Press, 1997.

———. *Robots in Space: Technology, Evolution, and Interplanetary Travel.* Baltimore: Johns Hopkins University Press, 2008.

Lavery, David. *Late for the Sky: The Mentality of the Space Age.* Carbondale: Southern Illinois University Press, 1992.

Law, John. *Power, Action, and Belief: A New Sociology of Knowledge?* London: Routledge & Kegan Paul, 1986.

Lee, Benjamin. Foreword to *Metaculture: How Culture Moves through the World,* by Greg Urban, ix–xvi. Minneapolis: University of Minnesota Press, 2001.

Levine, Caroline. *Forms: Whole, Rhythm, Hierarchy, Network.* Princeton, N.J.: Princeton University Press, 2015.

Lewis, John S. *Mining the Sky: Untold Riches from the Asteroids, Comets, and Planets.* Reading, Mass.: Addison-Wesley, 1996.

Limerick, Patricia Nelson. "The Adventures of the Frontier in the Twentieth Century." In *The Frontier in American Culture: An Exhibition at the Newberry Library, August 26, 1994–January 7, 1995,* essays by Richard White and Patricia Nelson Limerick, edited by James R. Grossman. Berkeley: University of California Press, 1994.

Little, Paul E. "Environments and Environmentalisms in Anthropological Research: Facing a New Millennium." *Annual Review of Anthropology* 28 (1999): 253–84.

Little, Peter C. *Toxic Town: IBM, Pollution, and Industrial Risks.* New York: New York University Press, 2014.

Lo, Martin. "The Interplanetary Superhighway and the Origins Program." Paper presented at the IEEE Aerospace Conference, Big Sky, Montana, March 9–16, 2002.

Lock, Margaret. "The Epigenome and Nature/Nurture Reunification: A Challenge for Anthropology." *Medical Anthropology* 32, no. 4 (2013): 291–308.

———. *Twice Dead: Organ Transplants and the Reinvention of Death.* Berkeley: University of California Press, 2002.

Logsdon, John M., moderator. *Legislative Origins of the National Aeronautics and Space Act of 1958: Proceedings of an Oral History Workshop, April 3, 1992.* Washington, D.C.: NASA, 1998.

Lovelock, James. *Gaia: A New Look at Life on Earth.* 1979. Reprint, Oxford: Oxford University Press, 2000.

———. *The Revenge of Gaia: Earth's Climate in Crisis and the Fate of Humanity.* New York: Basic Books, 2006.

Luke, Timothy W. "On Environmentality: Geo-power and Eco-knowledge in the Discourses of Contemporary Environmentalism." *Cultural Critique,* no. 31 (Autumn 1995): 57–81.

Marburger, John. Keynote address delivered at the Forty-Fourth Robert H. Goddard Memorial Symposium, Greenbelt, Maryland, March 15, 2006.

Marcus, George E., and Erkan Saka. "Assemblage." *Theory, Culture & Society* 23, nos. 2–3 (2006): 101–6.

Martin, Emily. *Flexible Bodies: Tracking Immunity in American Culture from the Days of Polio to the Age of AIDS.* Boston: Beacon Press, 1994.

Masco, Joseph. *The Nuclear Borderlands: The Manhattan Project in Post–Cold War New Mexico.* Princeton, N.J.: Princeton University Press, 2006.

———. *The Theater of Operations: National Security Affect from the Cold War to the War on Terror.* Durham, N.C.: Duke University Press, 2014.

Maturana, Humberto R., and Francisco J. Varela. *Autopoiesis and Cognition: The Realization of the Living.* Dordrecht: D. Reidel, 1980.

McCray, W. Patrick. *The Visioneers: How a Group of Elite Scientists Pursued Space Colonies, Nanotechnologies, and a Limitless Future.* Princeton, N.J.: Princeton University Press, 2013.

McDougall, Walter A. *The Heavens and the Earth: A Political History of the Space Age.* 1985. Reprint, Baltimore: Johns Hopkins University Press, 1997.

McGuirk, Kevin. "A. R. Ammons and the Whole Earth." *Cultural Critique,* no. 37 (Autumn 1997): 131–58.

McKay, Christopher P. "Planetary Ecosynthesis on Mars: Restoration Ecology and Environmental Ethics." Earth System Science Education Alliance, Institute for Global Environmental Strategies, December 2007. http://esseacourses.strategies.org.

McKay, Christopher P., and Margarita M. Marinova. "The Physics, Biology, and Environmental Ethics of Making Mars Habitable." *Astrobiology* 1, no. 1 (2001): 89–109.

McKibben, Bill. *Eaarth: A Survivor's Guide.* New York: Time Books, 2010.

McLean, Stuart. "Black Goo: Forceful Encounters with Matter in Europe's Muddy Margins." *Cultural Anthropology* 26, no. 4 (2011): 589–619.

Meadows, Donella H. *Thinking in Systems: A Primer.* Edited by Diana Wright. White River Junction, Vt.: Chelsea Green, 2008.

Mellor, Felicity. "Colliding Worlds: Asteroid Research and the Legitimization of War in Space." *Social Studies of Science* 37, no. 4 (2010): 499–531.

Melosi, Martin V., and Joseph A. Pratt, eds. *Energy Metropolis: An Environmental History of Houston and the Gulf Coast.* Pittsburgh: University of Pittsburgh Press, 2007.

Meltzer, Eve. *Systems We Have Loved: Conceptual Art, Affect, and the Antihumanist Turn.* Chicago: University of Chicago Press, 2013.

Meltzer, Michael. *When Biospheres Collide: A History of NASA's Planetary Protection Programs.* Washington, D.C.: NASA, 2010.

Mendell, Wendell W. "A Gateway for Human Exploration of Space? The Weak Stability Boundary." *Space Policy* 17, no. 1 (2001): 13–17.

Merchant, Carolyn. *The Death of Nature: Women, Ecology, and the Scientific Revolution.* New York: Harper & Row, 1989.

Merlan, Francesca. "Male–Female Separation and Forms of Society." *Cultural Anthropology* 7, no. 2 (1992): 169–93.

Messeri, Lisa. *Placing Outer Space: An Earthly Ethnography of Other Worlds.* Durham, N.C.: Duke University Press, 2016.

Miles, Ralph F., Jr. "Introduction." In *Systems Concepts: Lectures on Contemporary Approaches to Systems,* edited by Ralph F. Miles Jr., 1–11. New York: John Wiley, 1973.

Mills, C. Wright. *The Power Elite.* 1956. Reprint, New York: Oxford University Press, 2000.

Milman, Oliver. "Paris Climate Deal: Trump Says He Now Has an 'Open Mind' about Accord." *Guardian,* November 22, 2016. https://www.theguardian.com.

———. "Trump to Scrap NASA Climate Research in Crackdown on 'Politicized Science.'" *Guardian,* November 23, 2016. https://www.theguardian.com.

Mindell, David A. *Digital Apollo: Human and Machine in Spaceflight.* Cambridge: MIT Press, 2008.

Mitchell, Timothy. *Carbon Democracy: Political Power in the Age of Oil.* New York: Verso, 2011.

Mitman, Gregg. *Breathing Space: How Allergies Shape Our Lives and Landscapes.* New Haven, Conn.: Yale University Press, 2007.

Moeller, Hans-Georg. *Luhmann Explained: From Souls to Systems.* Chicago: Open Court, 2006.

Mol, Annemarie. *The Body Multiple: Ontology in Medical Practice.* Durham, N.C.: Duke University Press, 2002.

Morrison, David. "The Impact Hazard: Advanced NEO Surveys and Societal Responses." In *Comet/Asteroid Impacts and Human Society: An Interdisciplinary Approach,* edited by Peter T. Bobrowsky and Hans Rickman, 163–73. Berlin: Springer, 2007.

———, ed. *The Spaceguard Survey: Report of the NASA International Near-Earth-Object Detection Workshop.* Washington, D.C.: NASA, 1992.

Morsiani, Paola, and Trevor Smith, eds. *Andrea Zittel: Critical Space.* Munich: Prestel, 2005.

Mugerauer, Bob. "Maturana and Varela: From Autopoiesis to Systems Applications." In *Traditions of Systems Theory: Major Figures and Contemporary Developments,* edited by Darrell P. Arnold, 158–78. New York: Routledge, 2014.

Munn, Nancy D. "The Cultural Anthropology of Time: A Critical Essay." *Annual Review of Anthropology* 21 (1992): 93–123.

Murphy, Keith M. "Design and Anthropology." *Annual Review of Anthropology* 45 (2015): 433–49.

———. *Swedish Design: An Ethnography.* Ithaca, N.Y.: Cornell University Press, 2014.

Nash, Linda Lorraine. *Inescapable Ecologies: A History of Environment, Disease, and Knowledge.* Berkeley: University of California Press, 2006.

National Aeronautics and Space Administration. *Bioastronautics Roadmap: A Risk Reduction Strategy for Human Space Exploration.* Houston: NASA Johnson Space Center, 2005. https://ntrs.nasa.gov.

———. *Human Research Program Integrated Research Plan.* Houston: NASA Johnson Space Center, 2010. https://www.nasa.gov.

———. *Human Research Program: 2014 Fiscal Year Annual Report.* Houston: NASA Johnson Space Center, 2014. https://www.nasa.gov.

———. "Man-Systems Integration Standards." NASA-STD-3000, vol. 1. Revision B, July 1995. https://msis.jsc.nasa.gov.

———. *NASA FY 2017 Budget Request.* Fact sheet. Washington, D.C.: NASA, 2017. https://www.nasa.gov.

———. *NASA's Exploration Systems Architecture Study.* Washington, D.C.: NASA, 2005. https://www.nasa.gov.

———. "NASA's Next Prototype Spacesuit Has a Brand New Look, and It's All Thanks to You." Journey to Mars, April 30, 2014. https://www.nasa.gov.

———. "NASA Space Flight Human-System Standard." NASA-STD-3001, vol. 1, "Crew Health." Revision A, July 2014. https://www.nasa.gov/hhp/standards.

———. *Suited for Spacewalking: A Teacher's Guide with Activities for Technology Education, Mathematics, and Science.* Washington, D.C.: NASA, 1998. https://www.nasa.gov.

National Research Council. *A Framework for K–12 Science Education: Practices, Crosscutting Concepts, and Core Ideas.* Washington, D.C.: National Academies Press, 2012.

Newman, Dava J., Jeff Hoffman, Kristen Bethke, Christopher Carr, Nicole Jordan, Liang Sim, et al., *Astronaut Bio-Suit System for Exploration Class Missions: NIAC Phase II Final Report, August 2005.* Atlanta: NASA Institute for Advanced Concepts, 2005. http://www.niac.usra.edu.

Nisbet, James. *Ecologies, Environments, and Energy Systems in Art of the 1960s and 1970s.* Cambridge: MIT Press, 2014.

Nussbaum, Martha. "Objectification." In *The Philosophy of Sex: Contemporary Readings,* edited by Alan Soble and Nicholas Power, 381–419. Lanham, Md.: Rowman & Littlefield, 2002.

Nye, David E. *America as Second Creation: Technology and Narratives of New Beginnings.* Cambridge: MIT Press, 2003.

Odum, Eugene. *Ecological Vignettes: Ecological Approaches to Dealing with Human Predicaments.* Amsterdam: Harwood Academic, 1998.

Olson, Valerie A. "The Ecobiopolitics of Space Biomedicine." *Medical Anthropology* 29, no. 2 (2010): 170–93.

———. "NEOspace: The Solar System's Emerging Environmental History and Politics." In *New Natures: Joining Environmental History with Science and Technology Studies,* edited by Dolly Jørgenson, Finn Arne Jørgensen, and Sarah Pritchard, 195–211. Pittsburgh: Pittsburgh University Press, 2013.

———. “Political Ecology in the Extreme: Asteroid Activism and the Making of an Environmental Solar System.” *Anthropological Quarterly* 85, no. 4 (2012): 1027–44.

Olson, Valerie, and Lisa Messeri. “Beyond the Anthropocene: Un-earthing an Epoch.” *Environment and Society* 6 (2015): 28–47.

Ong, Aihwa, and Stephen J. Collier, eds. *Global Assemblages: Technology, Politics, and Ethics as Anthropological Problems.* Malden, Mass.: Blackwell, 2005.

Organisation for Economic Co-operation and Development. *The Space Economy at a Glance 2014.* Paris: OECD Publishing, 2014.

Ortner, Sherry B. *Sherpas through Their Rituals.* Cambridge: Cambridge University Press, 1978.

Paine, Thomas O. “Biospheres and Solar System Exploration.” In *Biological Life Support Systems: Commercial Opportunities,* edited by Mark Nelson. Washington, D.C.: NASA, 1990.

Palumbo-Liu, David, Bruce Robbins, and Nirvana Tanoukhi. *Immanuel Wallerstein and the Problem of the World: System, Scale, Culture.* Durham, N.C.: Duke University Press, 2011.

Paul, Richard, and Steven Moss. *We Could Not Fail: The First African Americans in the Space Program.* Austin: University of Texas Press, 2015.

Paxson, Heather. “Post-Pasteurian Cultures: The Microbiopolitics of Raw-Milk Cheese in the United States.” *Cultural Anthropology* 23, no. 1 (2008): 15–47.

Pedersen, Morten Axel. “The Fetish of Connectivity.” In *Objects and Materials,* edited by Penny Harvey et al., 197–207. London: Routledge, 2014.

Pendle, George. *Strange Angel: The Otherworldly Life of Rocket Scientist John Whiteside Parsons.* Orlando, Fla.: Harcourt, 2005.

Penley, Constance. *NASA/Trek: Popular Science and Sex in America.* New York: Verso, 1997.

Petryna, Adriana. *Life Exposed: Biological Citizens after Chernobyl.* Princeton, N.J.: Princeton University Press, 2002.

Phillips, John L. *The Bends: Compressed Air in the History of Science, Diving, and Engineering.* New Haven, Conn.: Yale University Press, 1998.

Pickering, William H. “Systems Engineering at the Jet Propulsion Laboratory.” In *Systems Concepts: Lectures on Contemporary Approaches to Systems,* edited by Ralph F. Miles Jr., 125–50. New York: John Wiley, 1973.

Pitts, Bradley, Cam Brensinger, Joseph Saleh, Chris Carr, Patricia Schmidt, and Dava Newman. *Astronaut Bio-Suit for Exploration Class Missions: NIAC Phase I Report, 2001.* Cambridge: MIT Man–Vehicle Lab, 2001.

Pitts, John A. *The Human Factor: Biomedicine in the Manned Space Program to 1980.* Washington, D.C.: NASA, 1985.

*Planetary.* Directed by Guy Reid. Planetary Collective, 2015. DVD, 85 min.

*Planetary Defense.* Directed by M. Moidel. Space Viz Productions, 2008. DVD, 48 min.

Povinelli, Elizabeth A. *Geontologies: A Requiem to Late Liberalism.* Durham, N.C.: Duke University Press, 2016.

Poynter, Jane. *The Human Experiment: Two Years and Twenty Minutes inside Biosphere 2.* New York: Thunder's Mouth Press, 2006.

Pritchard, Sara B. *Confluence: The Nature of Technology and the Remaking of the Rhône.* Cambridge, Mass.: Harvard University Press, 2011.

Rabinow, Paul. *Anthropos Today.* Princeton, N.J.: Princeton University Press, 2003.

———. *French Modern: Norms and Forms of the Social Environment.* Chicago: University of Chicago Press, 1995.

Rabinow, Paul, and Nikolas Rose. "Biopower Today." *BioSocieties* 1, no. 2 (2006): 195–217.

Ramo, Simon. "The Systems Approach: Wherein a Cure for Chaos Presents Itself." In *Systems Concepts: Lectures on Contemporary Approaches to Systems,* edited by Ralph F. Miles Jr., 13–32. New York: John Wiley, 1973.

Rancière, Jacques. *Disagreement: Politics and Philosophy.* Minneapolis: University of Minnesota Press, 1999.

Rasch, William, and Cary Wolfe, eds. "The Politics of Systems and Environments, Part I." Special issue. *Cultural Critique,* no. 30 (Spring 1995).

———, eds. "The Politics of Systems and Environments, Part II." Special issue. *Cultural Critique,* no. 31 (Autumn 1995).

Redfield, Peter. "The Half-Life of Empire in Outer Space." *Social Studies of Science* 32, nos. 5–6 (2002): 791–825.

———. *Space in the Tropics: From Convicts to Rockets in French Guiana.* Berkeley: University of California Press, 2000.

Reider, Rebecca. *Dreaming the Biosphere: The Theater of All Possibilities.* Albuquerque: University of New Mexico Press, 2009.

Review of U.S. Human Spaceflight Plans Committee. *Seeking a Human Spaceflight Program Worthy of a Great Nation.* Washington, D.C.: White House Office of Science and Technology Policy, 2009.

Rhatigan, Jennifer L., John B. Charles, and J. Michelle Edwards. "Exploration Health Risks: Probabilistic Risk Assessment." Paper presented at the International Astronautics Congress, Valencia, Spain, October 2006.

Rheinberger, Hans-Jörg. *Toward a History of Epistemic Things: Synthesizing Proteins in the Test Tube.* Stanford, Calif.: Stanford University Press, 1997.

Robinson, Kim Stanley, ed. *Future Primitive: The New Ecotopias.* New York: Tor, 1994.

Robinson, Michael F. *The Coldest Crucible: Arctic Exploration and American Culture.* Chicago: University of Chicago Press, 2006.

———. "The Explorer Gene." Time to Eat the Dogs (blog), September 28, 2009. https://timetoeatthedogs.com.

Rose, Nikolas. *The Politics of Life Itself: Biomedicine, Power, and Subjectivity in the Twenty-First Century.* Princeton, N.J.: Princeton University Press, 2007.

Rus, Mayer, and Charlotte M. Frieze. "Garrett Finney: The Thinker." *House & Garden,* June 2007.

Ruzic, Neil P. *Spinoff 1976: A Bicentennial Report.* Washington, D.C.: Technology Utilization Office, NASA, 1976.

Sadler, Simon. *Archigram: Architecture without Architecture.* Cambridge: MIT Press, 2005.

Said, Edward W. "Performance as an Extreme Occasion" (1989). In *The Edward Said Reader,* edited by Moustafa Bayoumi and Andrew Rubin, 317–46. New York: Vintage Books, 2000.

Sauvagnargues, Anne. "Crystals and Membranes: Individuation and Temporality." Translated by Jon Roffe. In *Gilbert Simondon: Being and Technology,* edited by Arne De Boever, Alex Murray, Jon Roffe, and Ashley Woodward, 57–70. Edinburgh: Edinburgh University Press, 2012.

Saxon, Dvera. "Strawberry Fields as Extreme Environments: The Ecobiopolitics of Farmworker Health." *Medical Anthropology* 34, no. 2 (2015): 166–83.

Schweickart, Russell L., Thomas D. Jones, Frans von der Dunk, and Sergio Camacho-Lara. *Asteroid Threats: A Call for Global Response.* Webster, Tex.: Committee on Near Earth Objects, Association of Space Explorers, September 25, 2008. http://www.space-explorers.org.

Shanken, Edward A., ed. *Systems.* Cambridge: MIT Press, 2015.

Shapin, Steven. *A Social History of Truth: Civility and Science in Seventeenth-Century England.* Chicago: University of Chicago Press, 1994.

Shapin, Steven, and Simon Schaffer. *Leviathan and the Air-Pump: Hobbes, Boyle, and the Experimental Life.* Princeton, N.J.: Princeton University Press, 2011.

Shapiro, Nicholas. "Attuning to the Chemosphere: Domestic Formaldehyde, Bodily Reasoning, and the Chemical Sublime." *Cultural Anthropology* 30, no. 3 (2015): 368–93.

Shaw, Debra Benita. "Bodies Out of This World: The Space Suit as Cultural Icon." *Science as Culture* 13, no. 1 (2004): 123–44.

Sherif, Maurice. *The American Wall: From the Pacific Ocean to the Gulf of Mexico.* Paris: MS Zephyr, 2011.

Shetterly, Margot Lee. *Hidden Figures: The American Dream and the Untold Story of the Black Women Mathematicians Who Helped Win the Space Race.* New York: William Morrow, 2016.

Shirley, Donna, with Danelle Morton. *Managing Martians.* New York: Broadway Books, 1998.

Shohat, Ella. "Notes on the 'Post-colonial.'" *Social Text,* nos. 31–32 (1992): 99–113.

Shostak, Sara. "Environmental Justice and Genomics: Acting on the Futures of Environmental Health." *Science as Culture* 13, no. 4 (2004): 539–62.

Siddiqi, Asif A. "Making Spaceflight Modern: A Cultural History of the World's First Space Advocacy Group." In *Societal Impact of Spaceflight,* edited by Steven J. Dick and Roger D. Launius, 513–37. Washington, D.C.: NASA, 2007.

Simondon, Gilbert. *On the Mode of Existence of Technical Objects.* Translated by Cécile Malaspina and John Rogove. Minneapolis: University of Minnesota Press, 2017.

Siskin, Clifford. *System: The Shaping of Modern Knowledge.* Cambridge: MIT Press, 2016.

Sloterdijk, Peter. *In the World Interior of Capital: Towards a Philosophical Theory of Globalization.* Translated by Wieland Hoban. Cambridge: Polity Press, 2013.

Smith, Robert W. *The Space Telescope: A Study of NASA, Science, Technology, and Politics.* Cambridge: Cambridge University Press, 1989.

Sobel, Dava. *The Planets.* New York: Viking, 2005.

Solan, Victoria Jane. "'Built for Health': American Architecture and the Healthy House, 1850–1930." PhD diss., Yale University, 2004.

*Space Odyssey: Voyage to the Planets.* Directed by Joe Ahearne. Impossible Pictures, 2004. DVD, 100 min.

SSP 2006 Team. *Luna Gaia: A Closed-Loop Habitat for the Moon—Final Report.* Strasbourg: International Space University, 2006. http://www.isunet.edu.

Stamatelatos, Michael. *Probabilistic Risk Assessment: What Is It and Why Is It Worth Performing It?* Washington, D.C.: NASA Office of Safety and Mission Assurance, 2000.

Star, Susan Leigh. "The Ethnography of Infrastructure." *American Behavioral Scientist* 43, no. 3 (1999): 377–91.

Star, Susan Leigh, and James R. Griesemer. "Institutional Ecology, 'Translations,' and Boundary Objects: Amateurs and Professionals in Berkeley's Museum of Vertebrate Zoology, 1907–39." *Social Studies of Science* 19, no. 3 (1989): 387–420.

Starosielski, Nicole. *The Undersea Network.* Durham, N.C.: Duke University Press, 2015.

Steltzner, Adam, and William Patrick. *The Right Kind of Crazy: A True Story of Teamwork, Leadership, and High-Stakes Innovation.* New York: Portfolio/Penguin Books, 2016.

Stengers, Isabelle. “The Cosmopolitical Proposal.” In *Making Things Public: Atmospheres of Democracy,* edited by Bruno Latour and Peter Weibel, 994–1003. Cambridge: MIT Press, 2005.

Stenner, Paul. “James and Whitehead: Assemblage and Systematization of a Deeply Empiricist Mosaic Philosophy.” *European Journal of Pragmatism and American Philosophy* 3, no. 1 (2011): 101–30.

Sténuit, Robert. *The Deepest Days.* Translated by Morris Kemp. New York: Coward-McCann, 1966.

Stepan, Nancy Leys. *“The Hour of Eugenics”: Race, Gender, and Nation in Latin America.* Ithaca, N.Y.: Cornell University Press, 1991.

———. *Picturing Tropical Nature.* Ithaca, N.Y.: Cornell University Press, 2001.

———. “Race and Gender: The Role of Analogy in Science.” *Isis* 77, no. 2 (1986): 261–77.

Stepaniak, Philip C., and Helen W. Lane, eds. *Loss of Signal: Aeromedical Lessons Learned from the STS-107* Columbia *Space Shuttle Mishap.* Washington, D.C.: NASA, 2014.

Stern, S. Alan. “Myriad Planets in Our Solar System and Copernicus Smiled.” SpaceDaily, September 11, 2006. http://www.spacedaily.com.

Stewart, Kathleen. *A Space on the Side of the Road: Cultural Poetics in an “Other” America.* Princeton, N.J.: Princeton University Press, 1996.

Stewart, Kathleen, and Susan Harding. “Bad Endings: American Apocalypsis.” *Annual Review of Anthropology* 28 (1999): 285–310.

Stone, Tanya Lee. *Almost Astronauts: The True Story of the “Mercury 13” Women.* Somerville, Mass.: Candlewick Press, 2009.

Strathern, Marilyn. *After Nature: English Kinship in the Late Twentieth Century.* Cambridge: Cambridge University Press, 1992.

———. *The Gender of the Gift: Problems with Women and Problems with Society in Melanesia.* Berkeley: University of California Press, 1988.

———. *Partial Connections.* Updated ed. Walnut Creek, Calif.: AltaMira Press, 2004.

———. “Reading Relations Backwards.” *Journal of the Royal Anthropological Institute* 20, no. 1 (2014): 3–19.

———. *The Relation: Issues in Complexity and Scale.* Cambridge: Prickly Pear Press, 1995.

Strughold, Hubertus. “General Review of Problems Common to the Fields of Astronomy and Biology.” *Publications of the Astronomical Society of the Pacific* 70, no. 412 (1958): 43–53.

Strydom, Piet. *Risk, Environment, and Society: Ongoing Debates, Current Issues, and Future Prospects.* Buckingham: Open University Press, 2002.

Suchman, Lucy A., Jeanette Blomberg, Julian E. Orr, and Randall Trigg. "Reconstructing Technologies as Social Practice." *American Behavioral Scientist* 43, no. 3 (1999): 392–408.

Suchman, Lucy A., Jeanette Blomberg, and Randall Trigg. "Working Artefacts: Ethnomethods of the Prototype." *British Journal of Sociology* 53, no. 2 (2002): 163–79.

Suedfeld, Peter. "Space Memoirs: Value Hierarchies before and after Missions—A Pilot Study." *Acta Astronautica* 58, no. 11 (2006): 583–86.

Swyngedouw, Erik. "Depoliticized Environments: The End of Nature, Climate Change and the Post-political Condition." *Royal Institute of Philosophy Supplement* 69 (2011): 253–74.

Taylor, Peter J. "Technocratic Optimism, H. T. Odum, and the Partial Transformation of Ecological Metaphor after World War II." *Journal of the History of Biology* 21, no. 2 (1988): 213–44.

Topham, Sean. *Where's My Space Age? The Rise and Fall of Futuristic Design.* Munich: Prestel, 2003.

Trachtenberg, Alan. *The Incorporation of America: Culture and Society in the Gilded Age.* 25th anniversary ed. New York: Hill and Wang, 2007.

Traweek, Sharon. *Beamtimes and Lifetimes: The World of High Energy Physicists.* Cambridge, Mass.: Harvard University Press, 1988.

Tresch, John. "Cosmogram." In *Cosmograms,* edited by Jean-Christophe Royoux and Melik Ohanian, 67–76. New York: Sternberg, 2005.

Tribbe, Matthew D. *No Requiem for the Space Age: The Apollo Moon Landings and American Culture.* Oxford: Oxford University Press, 2014.

Trompette, Pascale. "Revisiting the Notion of Boundary Object." Translated by Dominick Vinck. *Revue d'Anthropologie des Connaissances* 3, no. 1 (2009): 3–25.

Tsing, Anna Lowenhaupt. *The Mushroom at the End of the World: On the Possibility of Life in Capitalist Ruins.* Princeton, N.J.: Princeton University Press, 2015.

Tuan, Yi-fu. *Topophilia: A Study of Environmental Perception, Attitudes, and Values.* New York: Columbia University Press, 1990.

Turner, Fred. *From Counterculture to Cyberculture: Stewart Brand, the Whole Earth Network, and the Rise of Digital Utopianism.* Chicago: University of Chicago Press, 2006.

Urban, Greg. *Metaculture: How Culture Moves through the World.* Minneapolis: University of Minnesota Press, 2001.

U.S. Senate, Committee on Aeronautical and Space Sciences. *Project Mercury: Man-in-Space Program of the National Aeronautics and Space Administration.* 86th Congress, 1st Session, Report 1014. Washington, D.C.: Government Printing Office, December 1, 1959.

Valentine, David. "Atmosphere: Context, Detachment, and the View from above Earth." *American Ethnologist* 43, no. 3 (2016): 511–24.

Van de Vijver, Gertrudis, Linda Van Speybroeck, and Dani De Waele. "Epigenetics: A Challenge for Genetics, Evolution, and Development?" *Annals of the New York Academy of Sciences* 981 (December 2002): 1–6.

Vaughan, Diane. *The* Challenger *Launch Decision: Risky Technology, Culture, and Deviance at NASA.* Chicago: University of Chicago Press, 1996.

Venkatesan, Soumhya, Matei Candea, Casper Bruun Jensen, Morten Axel Pedersen, James Leach, and Gillian Evans. "The Task of Anthropology Is to Invent Relations: 2010 Meeting of the Group for Debates in Anthropological Theory." *Critique of Anthropology* 32, no. 1 (2012): 43–86.

Vertesi, Janet. *Seeing Like a Rover: How Robots, Teams, and Images Craft Knowledge of Mars.* Chicago: University of Chicago Press, 2012.

Vogler, Andreas. "The Universal House: An Outlook to Space-Age Housing." In *Concept House 1: Towards Customised Industrial Housing,* edited by Mick Eekhout, 77–87. Delft: Delft University of Technology, 2005.

Vogler, Andreas, and Arturo Vittori. "Space Architecture for the Mother Ship: Bringing It Home." In *Out of This World: The New Field of Space Architecture,* edited by A. Scott Howe and Brent Sherwood, 393–404. Reston, Va.: American Institute of Aeronautics and Astronautics, 2009.

von Bengtson, Kristian, and Evan Twyford. "Human Subsystem: Definition and Integration." In *AIAA Space 2007 Conference and Exposition, Long Beach, California, 18–20 September 2007.* Reston, Va.: American Institute of Aeronautics and Astronautics, 2007.

Von Schnitzler, Antina. "Traveling Technologies: Infrastructure, Ethical Regimes, and the Materiality of Politics in South Africa." *Cultural Anthropology* 28, no. 4 (2013): 670–93.

Waldman, Scott. "NASA Nominee Wants to Study Climate Change—on Mars." ClimateWire, September 19, 2017. https://www.eenews.net/cw.

Waligora, J. M., D. J. Horrigan, and A. Nocogossian. "The Physiology of Spacecraft and Space Suit Atmosphere Selection." *Acta Astronautica* 23 (1991): 171–77.

Ward, Peter Douglas. *The Medea Hypothesis: Is Life on Earth Ultimately Self-Destructive?* Princeton, N.J.: Princeton University Press, 2009.

Webb, George Ernest. "The Planet Mars and Science in Victorian America." *Journal of American Culture* 3, no. 4 (1980): 573–80.

Weddle, Perry. *Argument: A Guide to Critical Thinking.* New York: McGraw-Hill, 1978.

Weitekamp, Margaret A. *Right Stuff, Wrong Sex: America's First Women in Space Program.* Baltimore: Johns Hopkins University Press, 2004.

White, Frank. *The Overview Effect: Space Exploration and Human Evolution.* Boston: Houghton Mifflin, 1987.

White, Richard, and Patricia Nelson Limerick. *The Frontier in American Culture: An Exhibition at the Newberry Library, August 26, 1994–January 7, 1995.* Edited by James R. Grossman. Berkeley: University of California Press, 1994.

Whitehead, Ingrid. "Garrett Finney Brings Space Habitability Down to Earth." *Architectural Record,* April 2002.

Wilson, Albert G. "Problems Common to the Fields of Astronomy and Biology: A Symposium—Introduction." *Publications of the Astronomical Society of the Pacific* 70, no. 412 (1958): 41–43.

Wolf-Meyer, Matthew J. *The Slumbering Masses: Sleep, Medicine, and Modern American Life.* Minneapolis: University of Minnesota Press, 2012.

Wolf-Meyer, Matthew J., and Karen-Sue Taussig. "Extremities: Thresholds of Human Embodiment." *Medical Anthropology* 29, no. 2 (2010): 113–28.

Worster, Donald. *Nature's Economy: A History of Ecological Ideas.* 2nd ed. Cambridge: Cambridge University Press, 1994.

Wynter, Sylvia, and Katherine McKittrick. "Unparalleled Catastrophe for Our Species? Or, To Give Humanness a Different Future: Conversations." In *Sylvia Wynter: On Being Human as Praxis,* edited by Katherine McKittrick, 9–89. Durham, N.C.: Duke University Press, 2015.

Yeomans, Donald K. "The Torino Scale Shows We're Safe." In "The Lure of Rocks from Outer Space." *New York Times,* March 6, 2009. https://www.nytimes.com.

Young, Amanda. *Spacesuits: The Smithsonian National Air and Space Museum Collection.* Brooklyn, N.Y.: PowerHouse Books, 2009.

# INDEX

*Created by Eileen Quam. Page numbers in italics refer to illustrations.*

**Valerie Olson** is assistant professor of anthropology at the University of California, Irvine.